The Art of Astrophotography

In *The Art of Astrophotography*, astronomer and *Popular Astronomy* contributor Ian Morison provides the essential foundations of how to produce beautiful astronomical images. Every type of astroimaging is covered, from images of the Moon and planets, to the constellations, star clusters and nebulae within our Milky Way Galaxy and the faint light of distant galaxies. He achieves this through a series of worked examples and short project walk-throughs, detailing the equipment needed – starting with just a DSLR (digital single lens reflex) camera and tripod, and increasing in complexity as the book progresses – followed by the way to best capture the images and then how, step by step, these may be processed and enhanced to provide results that can rival those seen in astronomical magazines and books. Whether you are just getting into astrophotography or are already deeply involved, Morison's advice will help you capture and create enticing astronomical images.

Ian Morison is Emeritus Gresham Professor of Astronomy at the University of Manchester's Jodrell Bank Observatory. In addition to his academic credentials, he is a lifelong amateur astronomer, a founding member and patron of Macclesfield Astronomy Society and a past president of the UK's Society for Popular Astronomy, one of the country's largest amateur astronomy organisations. He has written four previous popular astronomy books, as well as an undergraduate astronomy textbook.

The Art of Astrophotography

Ian Morison
Jodrell Bank, University of Manchester

CAMBRIDGE
UNIVERSITY PRESS

University Printing House, Cambridge CB2 8BS, United Kingdom

One Liberty Plaza, 20th Floor, New York, NY 10006, USA

477 Williamstown Road, Port Melbourne, VIC 3207, Australia

314-321, 3rd Floor, Plot 3, Splendor Forum, Jasola District Centre, New Delhi - 110025, India

79 Anson Road, #06-04/06, Singapore 079906

Cambridge University Press is part of the University of Cambridge.

It furthers the University's mission by disseminating knowledge in the pursuit of education, learning and research at the highest international levels of excellence.

www.cambridge.org
Information on this title: www.cambridge.org/9781316618417

© Ian Morison 2017

This publication is in copyright. Subject to statutory exception and to the provisions of relevant collective licensing agreements, no reproduction of any part may take place without the written permission of Cambridge University Press.

First published 2017

A catalogue record for this publication is available from the British Library

Library of Congress Cataloging in Publication data
Names: Morison, Ian, 1943–
Title: The art of astrophotography / Ian Morison, Jodrell Bank [Centre for Astrophysics], University of Manchester.
Description: Cambridge : Cambridge University Press, 2017. |
Includes bibliographical references and index.
Identifiers: LCCN 2016047965 | ISBN 9781316618417 (pbk. : alk. paper)
Subjects: LCSH: Astronomical photography. | Solar system–Pictorial works.
Classification: LCC QB121 .M67 2017 | DDC 522/.63–dc23
LC record available at https://lccn.loc.gov/2016047965

ISBN 978-1-316-61841-7 Paperback

Cambridge University Press has no responsibility for the persistence or accuracy of URLs for external or third-party internet websites referred to in this publication, and does not guarantee that any content on such websites is, or will remain, accurate or appropriate.

This book is dedicated to my many friends at the Macclesfield Astronomical Society, which has been a most rewarding part of my life for over 25 years. Their encouragement has helped lead to this book.

Contents

Preface

The first point to address is why the title 'The Art of Astrophotography' was chosen for this book. As one finds when setting out to produce beautiful astronomical images, as much, or even more, time is spent processing the data as acquiring the raw data. There are many image processing tools available, some of which will be applicable to a particular image, and the appropriate ones need to be chosen and applied to give the desired result. Different images will require a different sequence of processes and rarely will these be the same, so there are no rigid rules as to how to go about it. There is thus some 'art' required to choose the appropriate tools to produce a particular image. But then, having produced an image, artistry is often required to make it aesthetically pleasing, so this is perhaps a further and more common use of the word 'art' as applied to astrophotography. Two examples of, hopefully, applying artistry in the production of an image are seen in the 'Star Trails' and 'Composite Meteor Trails' images in Chapters 1 and 17 respectively.

Software is extensively used not just for image acquisition but for processing the captured images. Happily, many of the required software tools are free, such as *Deep Sky Stacker* (DSS) and *Registax*, but specialist programs that are used to process astronomical data such as *Pixinsight* need to be purchased. One often finds that, having done the major part of the image processing in a specialist program, *Adobe Photoshop* is used to carry out the final editing of the image. The important point about *Photoshop* is that it can handle 16-bit data which less expensive programs such as *Photoshop Elements* cannot do. At the time of writing, the free program *GIMP* can also only handle 8-bit data, but it is said that the next version will be able to do so and thus this may then become an alternative to the use of *Photoshop.*

In this book all the image processing is handled in *Photoshop*. Adobe no longer sell the CS6 version (or any previous versions) of the program and it is leased on a month to month basis, so, at least, there is no large upfront cost. All of the capabilities required for the image processing techniques used in this book are found in the versions from CS2 onwards (I use an 'academic' version of CS4). It is sometimes possible to buy new copies of these older versions by searching on the Internet and, should you not have a copy of *Photoshop*, I would encourage you to search for '*Adobe Photoshop CS2*' to see what might be available.

The structure of the book is very simple. Relatively short chapters are used to illustrate virtually all aspects of astrophotography and are used essentially as 'worked examples', first describing the hardware that would be needed to capture a particular image and then describing, step by step, how the image is processed to produce the final result. As the chapters progress, the equipment required will become more complex, but they start with two chapters whose images require only the use of a DSLR (digital single lens reflex camera) and a tripod.

A few chapters introduce techniques that can aid all aspects of astroimaging, such as Chapter 10, which describes how DSLRs can be cooled to enhance their performance in summer, and Chapter 12, which shows how filters can be used to combat light pollution.

Gradually, all of the processing steps needed to create images are covered – in detail when first applied in the early chapters, and then with a summary of their use in later chapters. So even if one only wishes to use, for example, a telescope, equatorial mount and cooled CCD (charge-coupled device) camera, the earlier chapters should at least be read lightly.

The book's chapters are supplemented with a set of appendices to cover in more depth aspects of astroimaging that are briefly summarised within some of the chapters. Examples of appendices where introductory chapters are included in the main text are Appendices A and D relating to telescopes and autoguiding. A further appendix, Appendix B, discusses in some depth the mounts, alt/az or equatorial, that are such an important part of an imaging system.

The world's top astrophotgraphers tend to concentrate on one aspect of the hobby – and do it supremely well. Damian Peach's planetary images come to mind. Over the past few years I have been writing books and articles – most recently a series entitled 'Imaging for Beginners' for the UK magazine *Astronomy Now* – which have had to cover all aspects of astroimaging, so in one sense I am a 'jack of all trades' but master of none. (But, having said that, one of my lunar images has been awarded a trophy and a variety of others have been used in books and magazines to illustrate articles other than those of my own.)

I can only offer one piece of evidence that the advice given in this book can be useful to others. A friend of mine who is a superb natural history photographer asked me for advice as to how begin to take astroimages. I explained to him how I thought one could best image the Orion nebula region (which, as Chapter 7 explains, is not that trivial) and then process the set of images taken to produce a good final result. He submitted his image to the 'Insight Astronomy Photographer of the Year' competition organised by the Royal Observatory Greenwich and won the 2015 Sir Patrick Moore Prize for Best Newcomer award!

With a little application, anyone could easily achieve images similar to those shown in this book, and my sincere hope is that many of you who use it to learn about astroimaging will, in time, achieve images far better than any of those that I have made.

Acknowledgements

My friends at the Macclesfield Astronomical Society have been a great source of help and encouragement, and I offer particular thanks to Roy Sturmy for the loan of equipment and to Christopher Hill and Christopher Heapy for the use of images to help illustrate the book. Peter Shah and Damian Peach have also kindly allowed me to use their superb images. My thanks go also to them.

I would like to thank Vince Higgs, Lucy Edwards and the team at Cambridge University Press who have steered this book through to publication and to Linda Paulus at OOH Publishing who prepared this complex book for publication. Particular thanks go to Peter Gill, who copy-edits my articles for the magazine *Astronomy Now* and who has greatly helped me to improve my writing skills, and to Chris Cartwright, who has carried out a superb task in copy-editing what is a challenging and technical text.

Finally, but not least, I must thank my wife for supporting me through too many hours spent at the computer, and for graciously accepting the fact that, far too often, mounts and telescopes have been spread over the lounge and dining room ready for use and that one bedroom has been taken over for their storage.

1
Imaging Star Trails

In the days of film, capturing star trails used to be a relatively easy project. In the classic example, one simply pointed the camera at the Pole Star and took a single long exposure. With digital cameras, things are not quite so easy due to the 'dark current' produced in the sensor which greatly increases the noise level and overwhelms the image for very long exposure times. Thus the long exposure required to show star trails has to be made up of many short (~30 seconds) exposures which must then be combined to give the final image. This used to be a very complex business, but happily a wonderful program written by Marcus Enzweiler and called *StarStax* has become available to freely download and this takes away all the hard work. It may be downloaded from the Softpedia.com website – just search for 'StarStax Softpedia'.

But first one has to find a suitable location to take the multiple images. If one looks at similar star trails images on the Internet, those that stand out have interesting foregrounds – in one, the lamp housing of a lighthouse covers the Pole Star; in another, an attractive church lies in the foreground. So it is well worth trying to find such a location. To take the classic view of stars trailing around the North Celestial Pole one also needs an unobstructed view towards the north. I have often taken astronomy groups to a parking place on the south side of a mere (lake) near my home in Cheshire. This is about as dark a location as one can easily find in east Cheshire and has an open view to the north. I hoped that the lake would provide an interesting foreground but knew that there would be significant light pollution towards the north as one is looking over Manchester. I went there on a still evening with a transparent sky overhead (so reducing the effects of light pollution), equipped with a sturdy tripod on which to mount my Nikon D7000 camera and Sigma 10–20 mm lens. Wide angle lenses tend to give the most impressive views.

By the lakeside and alongside a convenient bench (as seen in Figure 1.1) I set up the camera pointing towards the Pole Star and set the Sigma 10–20 mm zoom lens to 10 mm (15 mm equivalent focal length) and at its full aperture of f/3.5. I sat beside the camera, wearing several layers of clothing and beneath a duvet cover (the air temperature was four degrees Celsius as I arrived and dropped to zero degrees Celsius as time went by) and manually took a series of 30-second exposures at ISO 800: stored as JPEGs to minimise the post-processing. To reduce the effect of any camera shake as

Figure 1.1 Looking north from the south side of Redesmere lake from where the star trails image was taken

the shutter was pressed, a delay of two seconds was used prior to taking the exposure. It is important that the 'long exposure noise reduction' function of the DSLR is switched off. This takes a second 'dark frame' with the shutter closed so that any amplifier glow or hot pixels can be removed by subtraction from the image. If used, there would thus be 30 second gaps between each exposure – absolutely not what is wanted when taking a star trails image. Every so often I checked the image displayed on the rear screen for signs of dewing. Cassiopeia was directly above and, when fully dark adapted, I could faintly make out the band of the Milky Way arching overhead. The imaging session ended after 50 minutes as the camera lens, perhaps not surprisingly, had dewed up!

As will be seen, 50 minutes was long enough to produce an attractive image. But should the total imaging time required be significantly longer, further steps need to be taken to both prevent the lens dewing up and provide a continuous power supply to the camera. To achieve the former is both cheap and easy. An old sock can simply have its toe removed so that it can be pulled over the lens and two hand warmers placed within it. Once activated, these will supply heat for many hours. A more sophisticated, but expensive, approach is to buy a 'LensMuff' produced by Kevin Adams in the USA (www.kadamsphoto.com). Be warned though, if importing them into the UK, one may well have to pay VAT (not excessive) but also a Post Office handling fee (which is). However, they can also be used to prevent dew forming on the objective of small refractors and so can make a useful and far less expensive alternative to using a dew strap with its associated power supply and controller.

Figure 1.2 A 'LensMuff' to hold two or three hand warmers which will prevent dewing up of the camera lens

To provide a continuous power supply to the camera one can purchase an AC power adapter kit that provides a dummy battery powered from an external AC supply. These are available for virtually all DSLR cameras. As these are mains powered there is a further problem when using one at a dark site such as would usually be chosen for a star trails image. A 12 volt battery will be required as well as a 12 to 220 volt DC to AC power inverter. Many astronomers use SkyWatcher Power Tanks, which are available with capacities of 7 or 17 Ah, but there are several alternatives, such as jump-start batteries, which will often have a 12 volt socket to take the plugs normally used by the inverters for their 12 volt supply. (A very similar setup is used when imaging meteor trails and is shown in Figure 17.1.)

There is one further very useful accessory that can be used so that manual initiation of each exposure is not required and that is an intervalometer. This will automatically initiate each exposure and not cause any camera shake. When, say, a 30-second exposure is set in the camera's manual mode, the interval between exposures would be set to 31 seconds and a long sequence of exposures programmed. The Viltrox MC Series 'Digital Timer Remote Controller' that I have used can initiate a sequence of up to 399 exposures. It is important that the unit has the correct connector for the camera to be used. Some cameras have built-in intervalometers but I have found these to be somewhat fiddly to use, especially in the dark, and much prefer to use an external unit. They are not expensive and can also be used as a remote shutter release when single exposures are to be taken.

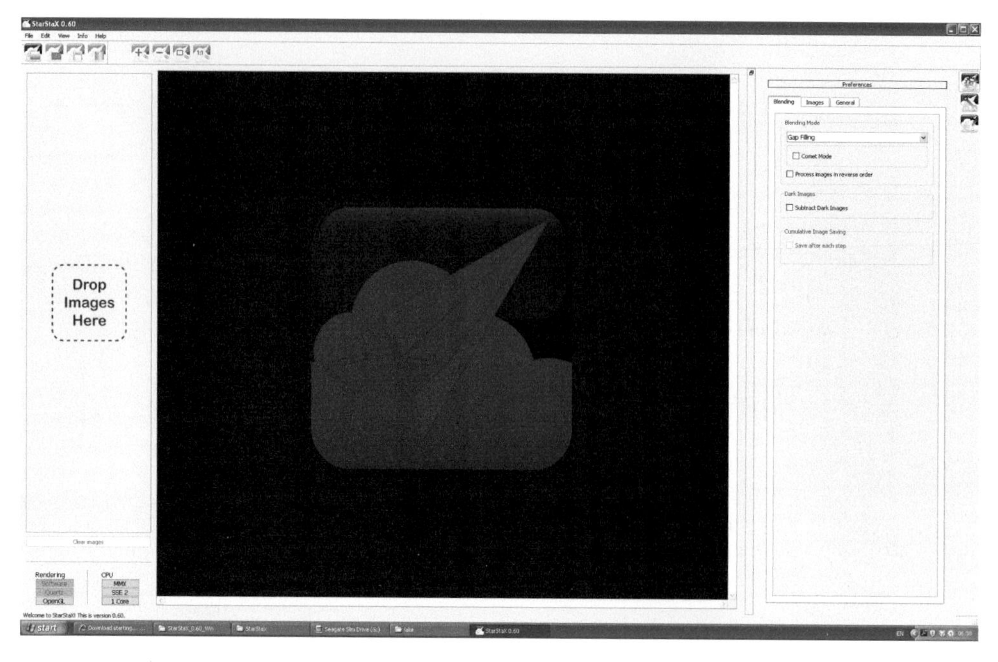

Figure 1.3 The *StarStax* opening screen

Back at home, the images were downloaded into a folder on the PC, and *StarStax* opened up as seen in Figure 1.3. The image processing was amazingly simple: first the 100 JPEG images were selected and dropped into the image box (seen on left in Figure 1.4) and then the 'Gap Filling' blending mode was selected in the preferences box, which was opened by clicking the 'gear wheel' icon at the top right of the screen. This mode fills in any gaps in the trails caused by the short breaks between exposures.

Though the images were heavily light polluted, the stars making up the Plough could be seen at the bottom of the first image and Polaris was very close to the image centre. The bright star at the upper left is Vega and the star just above the tree is Capella. The constellation Cassiopeia lies near the top of the frame.

The 'Start processing' icon (fourth from the left at the top left of the screen) was clicked and, as the program ran, the star trails gradually became apparent, taking a few minutes to build up the complete picture as seen, partly through the process, in Figure 1.5.

As I had seen in the camera rear screen display of individual images, the sky was an orange-red colour owing to light pollution from the Manchester area. The effect was not too unpleasant but more prominent than I wanted. I simply loaded the image produced by *StarStax* into *Adobe Photoshop*, opened the Levels box, selected the red level slider (rather than the overall RGB slider) and moved the 'black point' over to the right, so reducing the amount of red in the image. To avoid the railings in front of my camera, I had not included the lake when capturing the star trails, but to capture it, took a single 30-second exposure with the camera above the railings and tilted

Figure 1.4 The 100 JPEG images have been loaded into *StarStax,* which then shows the first image

Figure 1.5 *StarStax* has combined half of the 100 individual images

Figure 1.6 Star trails over Redesmere taken with a Nikon D7000 camera and Sigma 10 mm, f/3.5, wide angle lens

a little lower. *Photoshop* was then used to merge the star trails and lake images. The star trails and lake images were opened up together and the canvas size of the star trails image increased downwards using a black background. By opening the 'Clone Stamp' tool and suitably clicking, with the Alt key pressed, on a defined point in the lake image (this was a point at the far end of the lake), the star trails image was opened and the lake image cloned over the black extension on the star trails image to give the result seen in final image shown in Figure 1.6.

Though getting rather cold, sitting still under a dark and cloudless sky for nearly an hour was actually a rewarding experience and I felt that the resultant image had made it very worthwhile. I hope that some of you might try to achieve similar, and hopefully even better, results yourselves.

2 Imaging a Constellation with a DSLR and Tripod

This is a simple project for beginners but, as we will see, to get attractive results does require some effort in processing the images. The chapter introduces a highly important program called *Deep Sky Stacker* and a second, very useful, program called *IRIS*, and so is important even if images are only to be taken using telescopes.

The problem is that DSLR sensors are far better than film. When looking at a constellation image produced by a film camera, the brighter stars are perceived as such not so much by the fact that their images are brighter than others – all but the faintest stars will actually have a very similar brightness – but that their images are larger than those of fainter stars. This is similar to the way that star charts show stars of differing magnitudes. The reason is that film suffers from a problem called 'halation': the light from very bright objects can scatter off the back of the film and expose adjacent areas of the image, so making these regions, in our case bright stars, larger and more prominent. Though DSLR sensors do make the bright star images a little larger than the fainter ones, the effect is far less, and the initial image will not make the constellation patterns stand out. Digital constellation images can thus appear somewhat disappointing. Fear not – image processing programs such as *Adobe Photoshop* can be used to process the sky image to make the brighter stars more prominent and bring out their colours.

Aspects of the Use of DSLRs for Astroimaging

Sensor Size

Some years ago I would never have considered anything but a camera with an APSC (advanced photo system – C) sensor. At that time, the cost of full frame DSLRs having a sensor 36 × 24 mm in size was very high and for the same money one could purchase a cooled CCD camera which would be more worthwhile. However, prices have fallen considerably and it is now possible to purchase both Nikon and Canon full frame (FX) cameras for no more than the cost of my Nikon D7000 with an APSC sized sensor (DX) bought some years ago. The good thing about FX sensors is that they cover more than twice the area of sky as compared to a DX sensor (2.35 times in the case of Nikon APSC sensors and 2.63 times for Canon). This would make them very suitable for imaging meteor trails, as will be discussed in Chapter 17. It is not quite so obvious that they should be used with a telescope. Unless the telescope and field flattener elements (in telescopes termed astrographs) have been designed for use with an FX sensor, it is likely that images will suffer from both vignetting (darkening) and poor stellar images in the frame periphery. In the case of the former, discussed in Appendix E, corrections can be made by the use of flat fields (images of a uniform-brightness white surface) which are loaded along with the light frames (those of the object to be imaged) into the program to align and stack the images. In the case of the latter there are some techniques in *Photoshop* that can be used, which will be described in this and later chapters.

The Use of Raw and/or JPEG Files

A DSLR will typically read the data out from the sensor with a 12- or 14-bit analogue to digital converter (ADC). It may be possible to choose which bit depth is used, in which case choose the higher for astronomical imaging. If the output files are in the form of JPEGs only the most significant 8 bits will be output for each colour channel per pixel. As each pixel is only sensitive to one colour (as determined by the colour filter in the Bayer matrix above the sensor), interpolation is used to provide a colour value for the two colours per pixel that are not measured. The camera processor may well employ some scaling of the data before generating the JPEG file. There will thus be three bytes (8-bit words) saved per pixel. If a raw file is output, then for each pixel a 16-bit word will be saved. These should closely relate to the actual data readout from each pixel, but there may be some scaling. For example, the sensor in the Nikon D7000 is somewhat more sensitive to green light (as are most CCD or CMOS sensors) and so the values read out from the red and blue pixels are multiplied by 1.126 and 1.160 respectively before they are saved in the raw file. The software that processes the raw file has to know the format of the Bayer matrix used above the sensor in order to 'de-Bayer' the data to produce a colour image.

In general, therefore, it is better to save and process raw files, but as each image has to be passed through a raw converter, they can be slow to look through to find images, called light frames, that might suffer from tracking faults (perhaps due to a gust of wind) or have a plane or satellite passing across the field of view. I thus always save both raw and JPEG versions of each frame (though only JPEGs for star trail or meteor imaging as hundreds of frames might then be taken during an imaging session) as these can be easily scanned through to eliminate any poor frames before they are aligned and stacked.

Choice of ISO

An important difference between the ASA of a film and the ISO used by a digital camera when imaging is that a film with a high ASA (say 400) is actually more sensitive than one of low sensitivity (say 50). The silver halide crystals are larger in a high ASA film and so are more sensitive to light, but with the consequence that high ASA images will show more grain. In contrast, the sensitivity of a camera sensor is fixed and will depend on the size of the pixels and the technology employed. Modern camera sensors are far more sensitive than they were, partly because of the use of microlenses above each pixel which capture more light to focus into the pixel well, which is smaller than the overall size of the pixel owing to the supporting electronics surrounding the well.

When the ISO is increased, the gain of the amplifiers reading out the pixel data is increased, so making it appear that the sensitivity of the sensor is increased. This will tend to introduce more noise into the image, so high ISO images will naturally be noisier. The effect is often mitigated by the use of in-camera noise reduction – which is probably best turned off if JPEG files are to be used. Beyond some ISO value, which is thought to be 800 for my Nikon D7000 camera, the data readout (exactly as it would be at ISO 800) is simply multiplied by 2, 4, and 8 for ISOs of 1600, 3200 and 6400 respectively. As fewer effective bits are used, with the least significant bits containing more zeros as the ISO is increased, the noise level is bound to increase. So if raw files are to be saved – as will usually give the best results – there is little point in using ISOs greater than ~800.

When long exposures, of say 30 seconds, are made with a DSLR, the default mode is that 'long exposure noise reduction' will be applied. When the image (light frame) has been taken, a second exposure of the same length is taken with the shutter closed (a dark frame). The two images are then differenced in order to remove hot pixels and any amplifier glow that can sometimes be seen at the edge or corner of the image. This will obviously halve the effective exposure time, so it is probably best to turn off this camera mode. Hot pixels can easily be cloned out if they have not been removed by the stacking software. In general, the more photons detected from the sky, the better the final image.

Avoiding Camera Shake

Pressing the camera shutter will tend to cause a little tremor in the imaging system, which when imaging stars is surprisingly obvious if short exposures are being made. One can either use a remote release as discussed in Chapter 1 or employ a two second delay prior to each exposure. The lifting of the mirror in a DSLR can also cause a very slight tremor and a one second delay following its lifting can also be employed. However, if the camera is in 'Live View' mode, the mirror is always raised. (Live view is also an excellent aid to focusing.)

Taking a Constellation Image

First a suitable sky image must be taken and it is sensible to choose an area of the sky with some prominent stars making up one or more constellation patterns. For this chapter's examples, I chose to image the region encompassing Cassiopeia and Perseus from a dark sky site in Wales.

Choosing a Lens

The sensitivity of a camera to stars depends only on the effective area of the aperture of the chosen lens. As fully discussed in Chapter 17, that of a 50 mm, f/1.8 lens is several times greater than that of a 24 mm, f/2.5 lens. But, of course, the shorter focal length covers a greater area of sky. So a lens, preferably with a wide aperture (low f number) that will just cover the chosen area should be used. Prime lenses tend to give somewhat better image quality and have wider apertures than zoom lenses, though with the superb anti-reflection coatings now in use, the image quality difference is not as great as it once was. Most, but not all, lenses perform better when stopped down by one or two stops and the optimum f-stop to use can often be found by searching for lens reviews on the Internet.

Focusing

Most current DSLR cameras have a 'Live View' mode, which provides a continuous display of the scene being imaged, which is a very great help when focusing the lens. The camera must be in 'Manual Focus' mode (usually a switch beside the lens mount) and the focus initially set to the infinity mark on the lens. Using Live View, increase the magnification of the image whilst observing a bright star and make any fine adjustments if required. If the camera does not have Live View, one will have to take some test exposures and inspect the captured images at high magnification.

Imaging Cassiopeia and Perseus

I chose to use an excellent 40 mm equivalent, f/2.8 prime lens, stopped down to f/4. When using a fixed tripod, the Earth's rotation causes the stars to become elongated unless the exposures are very short. This effect is greatest when using a long focal length lens near the Celestial Equator and least when using a short focal length lens near the North Celestial Pole. So, when using a fixed tripod, it is best not to use a full frame equivalent focal length greater than ~50 mm (so 35 mm for an APSC sensor). With such a lens, exposures of 10 seconds will be about right for Orion at declination zero, on the Celestial Equator. Imaging nearer the pole and using shorter focal lengths will allow longer exposures: up to about 30 seconds when using a 24 mm equivalent lens to image Ursa Major. To acquire good images it is important to use a sturdy tripod, then select a mid-range ISO, say 800, and give a 2-second delay between triggering the camera and taking the exposure (or use a remote release such as an intervalometer, as described in

Chapter 1). Raw images will give the best results but, as mentioned above, I always take JPEGS as well, so I can easily inspect them to check for any problems.

The technique is to take 20 or so short exposures, called light frames, and then use the free program, *Deep Sky Stacker*, to combine them into a deeper image. This is available for PCs and can be downloaded from http://deepskystacker.free.fr/english/index.html. It is possible to run *Deep Sky Stacker* on an Apple Mac by putting it in a 'wrapper', as discussed in http://blog.tom-goetz.org/2013/01/running-deep-skystacker-for-windows-on.html. This provides a link to enable *Deep Sky Stacker* to become a Mac application. If *Deep Sky Stacker* will not accept the raw files from your camera, *Adobe DNG Converter* will convert them to .dng files which DSS will read. Again, this can be downloaded for free from www.adobe.com/support/downloads/product.jsp?product=106&platform=Windows. There are many tutorials on using *Deep Sky Stacker* on the Internet, such as www.stevebb.com/deep_sky_stacker.html, and as this program will be used in many of the imaging examples, these could be read to support the details given in this book.

Using *Deep Sky Stacker* to Align and Stack the 'Light' Frames Taken of the Constellations

Having opened *Deep Sky Stacker*, the exposures are loaded by clicking on 'Open picture files' at the top left of the screen, when a box appears headed 'Open light frames'. The file containing the images is opened and the images selected. The next step is to click on 'Check all' when a tick appears to the left of each image filename.

When 'Stack checked pictures' is clicked upon, *Deep Sky Stacker* will find the positions of stars in the frames and then measure their offsets in both position and angle with a reference frame. This can cause problems and it is quite a good idea to first just select two frames to stack. As *Deep Sky Stacker* works through them, one can see how many stars are found in the 'Registering stars' box that appears. If one sees that well over a thousand stars are found, the alignment process will take quite a long time or, as happens more often, one can find that *Deep Sky Stacker* cannot find sufficient stars to align on and finally reports that it will only stack one frame. In either case, one needs to click on 'Register checked pictures' and click on the 'Advanced' tab within the box that appears. As seen in Figure 2.1, a slider adjusts the star detection threshold. This should be increased if too many stars are being detected (100 to 200 star are quite sufficient) or reduced if *Deep Sky Stacker* cannot find sufficient. The reason that I think that it is worth trying with just two frames is that if one finds that too many stars are being detected, the alignment process can take a very long time, and although there is a 'Stop' button in the registering box, I have never found this to work.

Once with a particularly challenging set of light frames, I found that insufficient stars could be detected even with the star threshold set to its minimum value of 2 per cent. There is only one solution to this problem and that is to individually 'stretch' each frame prior to stacking. Each image is loaded into *Adobe Photoshop* and, using the *Levels* command (Image > Adjustments > Levels), the centre slider is moved to the left to give a value of ~1.20. This brightens the fainter parts of the image, so making fainter

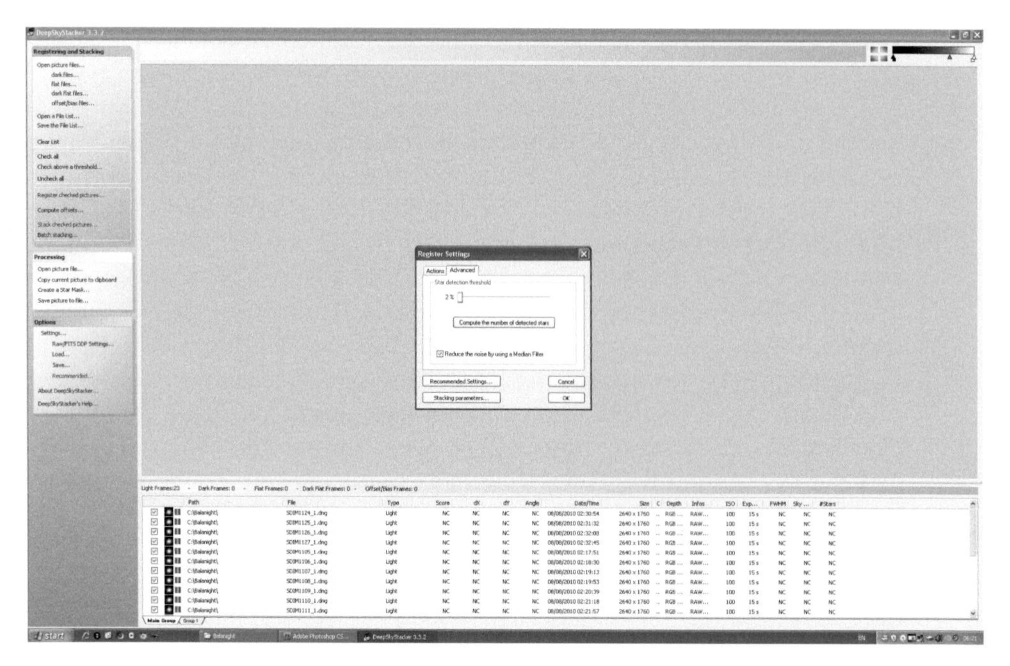

Figure 2.1 The individual short exposures are loaded into *Deep Sky Stacker* and the star detection threshold adjusted down to 2 per cent

stars more prominent. It should be made an 'Action' (i.e. saved for future use – see the following section) so that identical stretching is applied to all frames.

Making a *Photoshop* Action

The 'Action' window is opened by ticking on 'Window' in the list of options at the top and checking the 'Actions' box. At its top right is a small box with a downwards arrow and several lines to open up the Actions menu. 'New action' is clicked upon and a window opens up to enable one to select a function key to initiate the action. Having selected this, one clicks on 'Record' and then the required sequence of commands is carried out. Once complete, the Actions menu window is opened again and 'Stop recording' clicked upon. The chosen function key will then carry out the sequence of commands with a single press. This stretching process might well have to be applied twice to each frame before loading them into DSS for a second (and usually successful) attempt. This is one of several methods available to stretch an image and so bring up the fainter parts of the image.

Stretching the Output Given by *Deep Sky Stacker*

As is always found, the TIFF image produced by *Deep Sky Stacker* is very disappointing and, as seen in Figure 2.2, shows very few stars. The DSS image has then to be

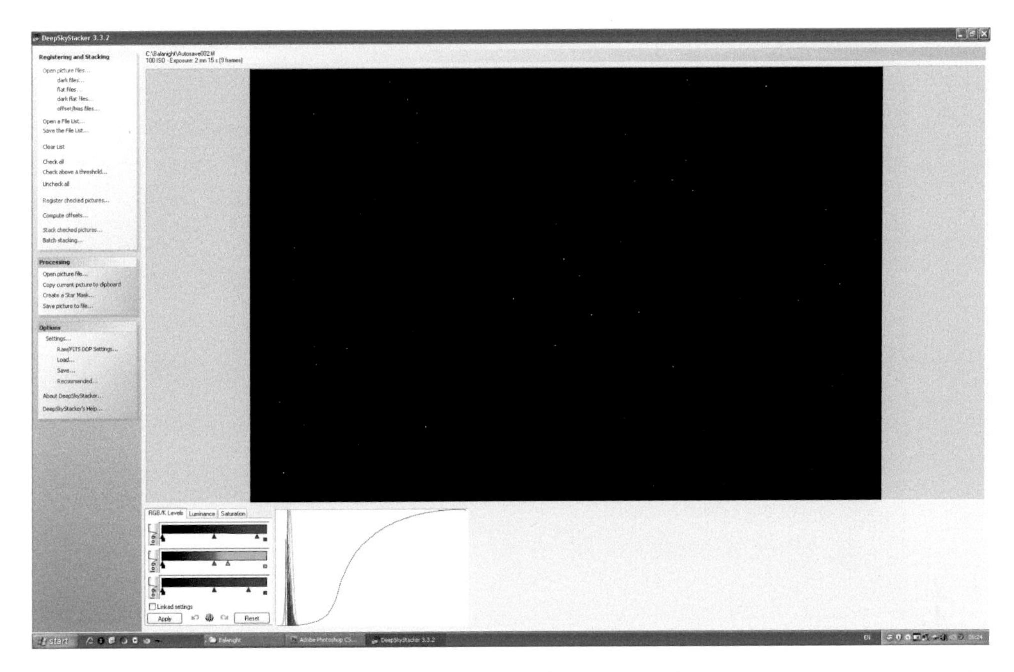

Figure 2.2 The *Deep Sky Stacker* output showing, typically, very few stars due to the 16-bit image depth

'stretched' to bring up the fainter stars. The stretching can be applied using *Photoshop* as described above, but can also be achieved using the free program *IRIS*, unfortunately only available for PCs, which can be downloaded from www.astrosurf.com/buil/us/iris/iris.htm. When opening up the image (using the load command) a stretched image will immediately appear, but may well be improved by selecting 'Logarithm' in the 'View' drop-down menu and adjusting the two sliders that are provided. In the case of this constellation image, the initial stretching was perfect and could be saved. At this point the stars in the constellations were visible, but the bright stars making up the constellation patterns were not very prominent, as can be seen in Figure 2.3.

Correcting for any Slight Star Trailing

When using a fixed tripod, even with an appropriate exposure time, some slight star trailing will usually be found, making stars look like very short sausages. Provided the effect is small, *Photoshop* can be used to make the stars round. The image is opened up and a duplicate layer made (Layer > Duplicate Layer > OK) or, more simply, using 'Ctrl-J'. The blending mode is set to 'Darken' and the 'Move' tool at the top of the tool bar is clicked upon, followed by a click on the image. The arrow keys may then be used to adjust the position of the top layer over the bottom layer and you will find that this makes the stellar images more rounded. When a good result has been achieved, the two layers are flattened (Layer > Flatten Image). Sometimes the effect of a single keystroke of the arrow keys is too much but there is a simple solution: increase the image

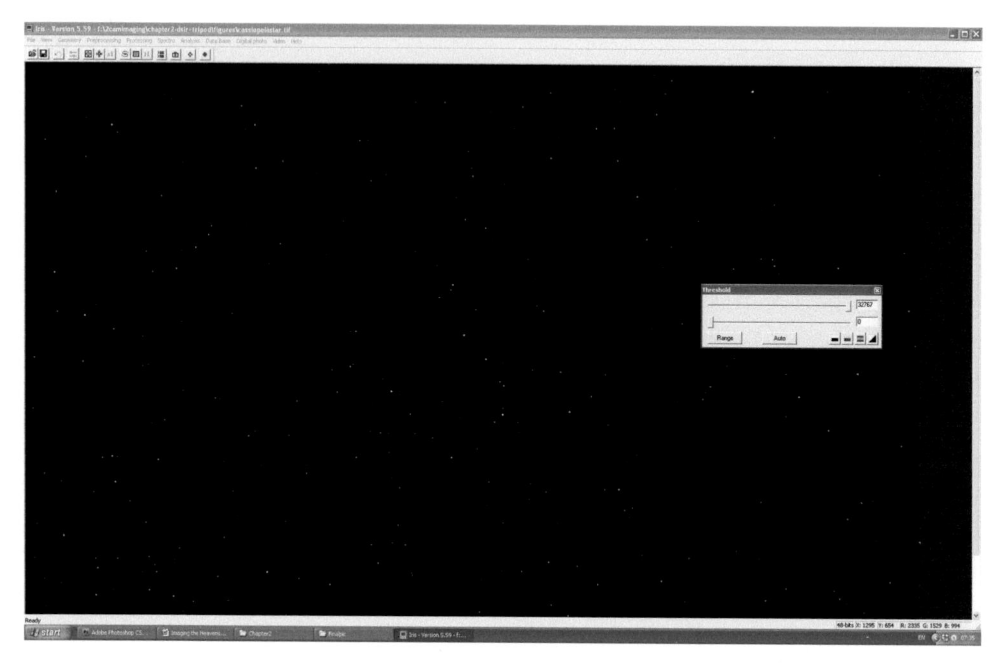

Figure 2.3 Having loaded the image produced by *Deep Sky Stacker* into the IRIS program immediately shows a stretched image – but the brighter stars do not stand out

size by a factor of two (Image > Size > 200 per cent), apply the correction, flatten the two layers and then reduce the size back to its original size (Image > Size > 50 per cent).

Enhancing the Stellar Images

The renowned Japanese astrophotographer Akira Fujii produced wonderful film based images of the constellations used to illustrate the book *The Great Atlas of the Stars*.[1] He used a diffusing (soft focus) filter (Hutech Scientific offer the Kenko 'Softon' filter) to expand the sizes of the brighter stars. This also helps to bring out the star colours, which, for bright stars, tend to saturate at pure white. It is also possible to achieve a similar result using *Photoshop*. First one has to select the brightest stars in the field. The image is loaded into *Photoshop*, and these are individually selected using the following process. One of these stars is selected using the 'Elliptical Marquee' Tool, which can be found by right clicking on the 'Rectangular Marquee' tool, second from the top in the tools menu. The ellipse is used to form a circle around the star using a feather of one pixel. The selection is made into a mask by clicking on the bottom item in the Tools menu (which appears as a rectangle with a circle within it). All but the selected star will be covered with a red, partially opaque, mask. The erasure tool is selected and, with a suitable diameter selected by use of the '[' and ']' keys, the other bright stars are removed from the mask (the mask is semi-opaque, so the brighter stars can be seen through it) to give

[1] *The Great Atlas of the Stars*, Serge Brunier and Akira Fujii. Firefly Books. ISBN 978-1-55209-610-6.

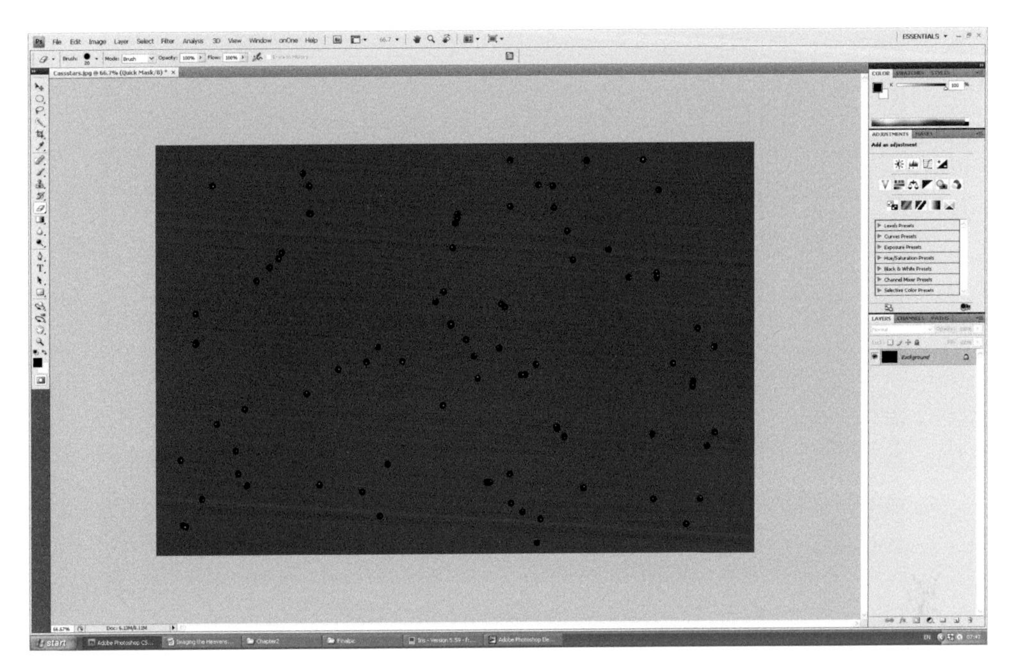

Figure 2.4 Erasing a mask to select the brightest stars

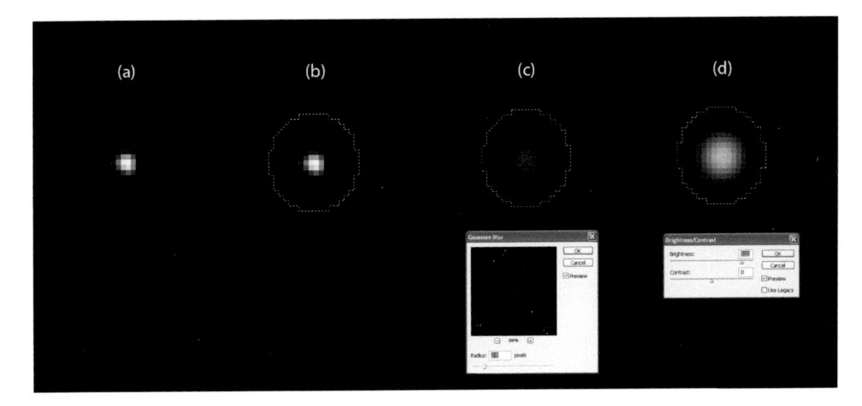

Figure 2.5 Enhancing the brighter stars as described in the text

a result as seen in Figure 2.4, which can then be converted back to the standard selection display of 'marching ants' surrounding the selected stars. As shown in the sequence of Figure 2.5 showing the effect on one bright star (Figure 2.5a) which is selected (Figure 2.5b), and the Gaussian Blur filter is applied with a radius of a few pixels (Filter > Blur > Gaussian Blur) to smooth and enlarge their images as seen in Figure 2.5c. This makes them somewhat fainter, and their brightness can be brought back by adjusting the 'brightness' slide within the Brightness/Contrast controls (Image > Adjustments > Brightness/Contrast) to give the result shown in Figure 2.5d. One very good aspect of using this technique is that it helps the star colours to become more obvious.

Figure 2.6 The final result showing the Cassiopeia constellation

It may be that one would like to increase the emphasis on the stars making up the constellation pattern to make them stand out even better. These relatively few stars are selected as above and further blurring and brightening processes applied to give the result shown in Figure 2.6 showing the Cassiopeia constellation. Figure 2.7 shows both the Cassiopeia and Perseus constellations taken at the same time and processed in the same way.

Figure 2.7 The Cassiopeia and Perseus constellations imaged with a DSLR mounted on a fixed tripod, derived from 20- and 15-second exposures taken with a 40 mm full frame equivalent, f/2.8 lens

3
Imaging the Milky Way with a DSLR and Tracking Mount

As we saw in the previous chapter, a DSLR with a fixed mount can be used to image the heavens, but the exposure times and lens focal lengths are limited by the Earth's rotation. These limitations can be removed if a tracking mount is used and this is particularly helpful when imaging the Milky Way as longer exposures do help to bring out its faint nebulosity. However, as light pollution will limit the faintest details that can be imaged, it is good to get to a really dark sky site and so keen astrophotographers often travel by air to dark locations across the world. In this case it is not feasible to take standard equatorial mounts and so a number of manufactures have brought out lightweight and compact equatorial heads that can be mounted on a camera tripod.

Tracking Mounts

Two camera sized equatorial heads that are currently available are the Vixen 'Polarie Star Tracker' and the iOptron 'Sky Tracker Pro'. The Vixen unit incorporates a small 'North Star alignment window' to polar align their mount (a polar scope is available as an accessory), while the iOptron unit has an integral rotating azimuth base and latitude adjustment wedge as well as an illuminated polar scope. Both will support cameras and lenses weighing up to a total of ~3 kg. Both use a standard worm and gear to drive the right ascension axis. In the case of the Polarie, a 9.0 mm brass worm drives a 57.6 mm diameter, 144 teeth aluminium worm wheel, while the Sky Tracker uses a 11 mm worm and 80 mm diameter, 156 teeth worm wheel. One nice feature of the Sky Tracker Pro is that it includes a precision, worm driven latitude adjuster so it can be mounted directly on the centre column of a tripod. The iOptron's included polar-alignment scope has an illuminated reticule with calibrated circles for Polaris (Northern Hemisphere) and Sigma Octantis (Southern). The circles are divided into 12 hours like the face of a clock. iOptron have created an app for Apple devices that calculates the correct angular position for the polar scope.

SkyWatcher has brought out the 'Star Adventurer', which is, in effect, a miniature equatorial mount that can support a camera system (or small telescope) up to 5 kg.

Figure 3.1 The Vixen Polarie Star Tracker, iOptron Sky Tracker Pro and SkyWatcher Star Adventurer tracking mounts

The mount includes an integral illuminated polar scope mounted within the polar axis. The equatorial head is mounted on an included equatorial wedge to screw onto a tripod bush and set for the observer's latitude. A counterweight shaft with a 1kg counterweight is employed to balance the system. A dovetail L-bracket called the 'Fine Tuning Mounting Assembly' fits on the equatorial head so one, or even two, cameras can be mounted on it using tripod ball heads. It can be powered by four integral AA batteries or by using a mini-B type USB port. The right ascension drive uses a 13 mm brass worm and 86 mm, 144 teeth aluminium worm wheel gear, so provides very similar tracking accuracy to the Vixen and iOptron mounts.

The tracking accuracy of these equatorial heads is limited by the size of the right ascension gear (around three inches in diameter) and will allow unguided exposures for up to 5 minutes when a short focal length lens is employed – just what is needed to image the arch of the Milky Way across the sky. Tests of these mounts have shown that the peak-to-peak tracking error is of the order of 25–30 arc seconds over the ~10 minute period which corresponds to one rotation of the worm. If one were using an 18 Mpixel sensor APSC camera, the pixel size is 4.3 microns in size. So, when using a 25 mm lens, each pixel corresponds to 35 arc seconds in angular diameter. If the mounts were perfectly aligned on the North Celestial Pole, one could, in principle make unlimited length exposures. This would be virtually the case with a 50 mm lens as well with a drift of only 2 pixels. Using longer lenses, exposure times would ideally be limited and some experimentation would be necessary. Any worm and gear drive can occasionally give 'glitches' which would affect some exposures. Exposure times of 30 seconds would probably be a suitable exposure time, with the individual frames inspected before the majority being stacked in, for example, *Deep Sky Stacker*. The

Figure 3.2 The Baader Nano Tracker tracking mount with an aluminium plate laser pointer mount to use for polar alignment. Hanging below is its battery pack and control unit

length of allowed exposure time does, of course, depend on how well the mounts have been polar aligned.

The smallest and lightest tracking mount (weighing just 360 grams) is the Baader Planetarium (or Sightron) Nano Tracker, which is also the least expensive. With a smaller right ascension gear of ~50 mm diameter it will, in principle, have a somewhat lower tracking accuracy than those described above. One weakness is that only a small, and not very practical, sight hole is provided for polar alignment. I have acquired one for use when travelling by air with a very limited luggage allowance and have made a simple aluminium plate on which I can mount a laser pointer for polar alignment, as seen in Figure 3.2. The laser beam must be offset from the Pole Star by 0.7 degrees (43 arc minutes to be precise – and about the width of a little finger at arm's length) towards the star Kochab in Ursa Minor.

Figure 3.3 shows the position of the North Celestial Pole relative to Polaris and Kochab. The direction of Kochab from Polaris is ~15 degrees in a clockwise direction around from the star Alkaid, which lies at the end of the Plough's handle and may be easier to spot. (The Plough is called the Big Dipper in the USA.) The pointing accuracy obtained by this method is perfectly adequate for the 30–60-second exposures that I employ, before using *Deep Sky Stacker* to give the equivalent of longer exposures.

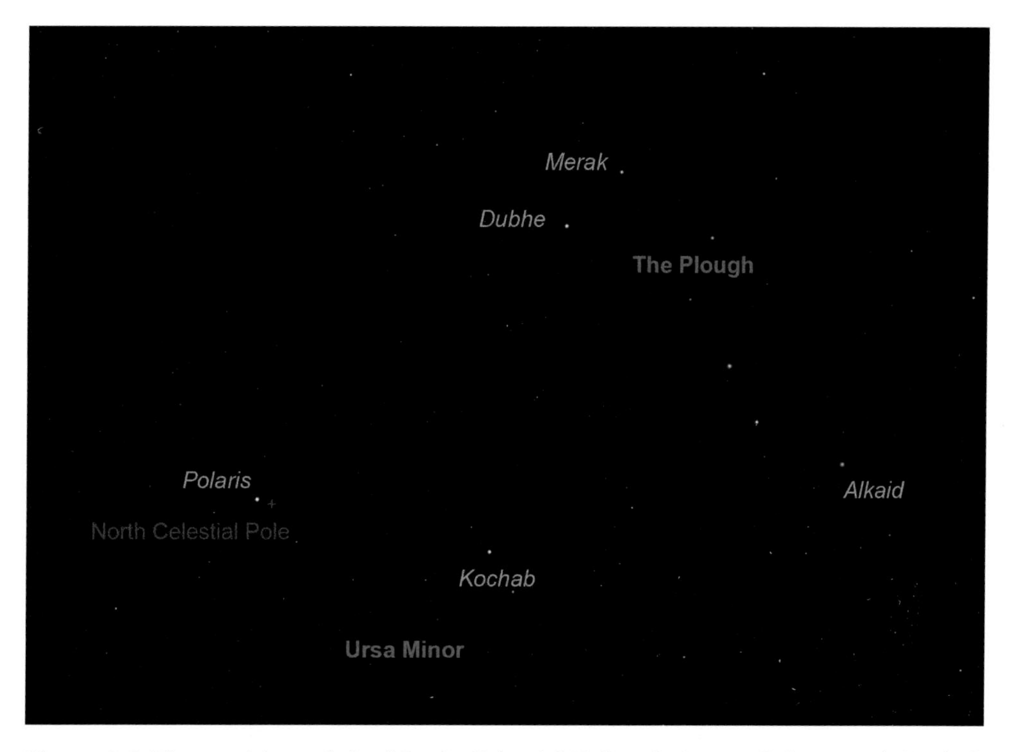

Figure 3.3 The position of the North Celestial Pole relative to Polaris and Kochab (Beta Ursa Minoris)

An innovative tracking mount is produced by the UK company AstroTrac, called the TT320X-AG. The 'AG' indicates that it is equipped with a ST4 autoguiding port to allow guiding in right ascension though I suspect that this facility will be rarely used. Once polar aligned, it can track accurately for nearly 2 hours. However, even without guiding it is capable of making exposures of up to 5 minutes with a typical tracking error of +/− 5 arc seconds (assuming accurate polar alignment). This accuracy is due to the fact that the drive, resulting from the rotation of a precision screw under the control of a small computer, is at the base of a long arm. Thus the angular motion of the mounting head is far less than the motion of the screw and the result is that any periodic error in the screw thread drive is greatly reduced. To achieve equivalent tracking accuracy a very expensive equatorial mount would be required. Though the mount weighs just 1 kg (2.2 pounds), given a suitably solid tripod or pillar, it can support up to 15 kg (33 pounds), so even 80 mm refractors or a Celestron C6 Schmidt–Cassegrain could be used. AstroTrac provide a TH3010 counterbalanced head to allow such telescopes to be mounted on it. A sturdy tripod is required with a pan and tilt head to mount the AstroTrac so that it can be polar aligned (the extendable bracket provided in which the illuminated polar mount is mounted is seen in Figure 3.4), along with a sturdy ball and socket head to hold the camera. They also provide a very neat and compact equatorial wedge to replace the pan and tilt head

Figure 3.4 The author's AstroTrac TT320X-AG with Nikon D7000 DSLR mounted on a pan and tilt head to enable polar alignment to be carried out using the illuminated polar scope

and an ultra-light pier which, for transporting by car or air, can contain and protect the AstroTrac itself. My AstroTrac, mounted on a Manfrotto tripod is shown in Figure 3.4. AstroTrac provide an optional illuminated polar scope to carry out the polar alignment which fits into, and is then held magnetically into, an extending arm. As an alternative, I have an adapted laser pointer that fits into the arm for polar alignment. I mount my Astrotrak on a Manfrotto 190XPROB professional tripod, initially using a solid pan and tilt head, but more recently the Manfrotto 410 Junior Geared Head to enable it to be polar aligned.

A further tracking mount which claims a peak-to-peak unguided tracking error of 2 arc seconds or less (assuming accurate polar alignment) is the FORNAX 10 LighTrack Mark II. Shown in Figure 3.5, it has been used successfully with lenses of focal length up to 600 mm. Like the AstroTrac, it uses a radius arm drive, but this is friction

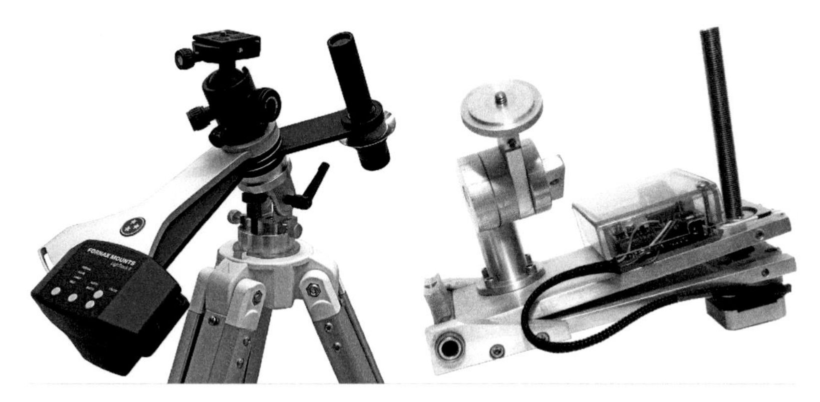

Figure 3.5 The 'FORNAX 10 LighTrack Mark II' and 'StarSync tracker' tracking mounts

rather than screw driven. A polarscope adapter ring that allows the use of a Celestron Polar Finder is included – or one could mount a laser pointer as I have done for use with the Nano Tracker and AstroTrac. A single tracking period is limited to around two hours.

A recent addition to the range of star trackers is the Star Sync tracker which (in a way I rather like) puts 'function-over-form'. It is a stainless steel and aluminium development of the simple 'barn door' tracking mount that many astrophotographers constructed and used before the current sophisticated star trackers were available. As also seen in Figure 3.5, the upper plate is driven by a stepper motor rotating a solid metal screw. The variable rate at which the screw needs to turn is controlled by a small 'Arduino' microcomputer to compensate for the fact that the screw is straight, not curved. It can track for over 2 hours before it resets automatically. The camera mounting head is built in, so a ball and socket joint is not required, and there is a channel in which to lay a laser pointer with which to align onto the North Celestial Pole. At the time of writing it is only available in the USA, but there should soon be UK and European distributors. As with the Astrotrac and Fornax LighTrak, the greater effective gear radius than the gears used in the compact star trackers should allow for more accurate tracking and the ability to take longer exposures. This is a very solidly made mount and thus could easily be used with longer focal length lenses than the smaller tracking mounts. When I was able to use one, I particularly liked the tall camera mounting head which allowed the camera to be aligned on the sky somewhat more easily than when using a ball and socket joint.

Colour Mottling

The renowned American astrophotographer Tony Hallas has highlighted a problem when using DSLRs for astroimaging. When a single image is highly stretched, the pattern of noise is very 'blotchy' and shows small areas of different colours, as seen in Figure 3.6, left. This is a very small area surrounding the Pleiades cluster taken from one frame that was used to provide the 'starfield' image provided for the Geminid

Figure 3.6 A demonstration of colour mottling, as described in the text

meteors as seen in Figure 17.5. Tony Hallas calls this chromatic noise 'colour mottling' and the way that he suggests to overcome it is to offset each frame randomly by perhaps up to a few star diameters so that, when stacked, the colours average out to a dark grey. Figure 3.6, right is the result of identically stretching the result of stacking 24 exposures all slightly offset from each other. Far less colour mottling is visible. The sequence of frames was taken with a camera mounted on the Baader Nano Tracker. As the alignment on the North Celestial Pole was not perfect, the sky image will have moved across the sensor as the sequence of frames was taken and this will also help to average out the colour noise. In fact, the tracking had been pretty good and the final frame in the stack had only shifted by 7 pixels in the X direction and 5 pixels in the Y direction from the first, but this was still sufficient to average out much of the mottling. The moral here is that perhaps it is best *not* to perfectly align on the North Celestial Pole as long as no trailing of stars is visible in a single frame. *Deep Sky Stacker* will align the stars in the sequence of frames and the only downside is that very slight cropping of the stacked result will be required.

Imaging the Heart of the Milky Way

In 2013, I was able to image the heart of the Milky Way in the constellations Sagittarius and Scorpius from a dark sky site in New Zealand. A total of 19 raw exposures, each of 30 seconds were taken at an ISO of 800 using a DSLR and a high quality, f/1.7, 40 mm equivalent prime lens used at f/2.8. These were stacked in *Deep Sky Stacker* and then stretched in *IRIS* (as described in Chapter 2) to produce the result seen in the Figure 3.7.

There is a problem that is found when trying to enhance nebula images such as the Milky Way or North America Nebula as often the stars then become too prominent. The solution, given *Adobe Photoshop*, is to first remove the stars from the image,

Figure 3.7 The stretched image of the Milky Way produced by the *IRIS* program from individual frames stacked in *Deep Sky Stacker*

then enhance the nebulosity and finally bring the stars back into the image with a suitable brightness so that they do not dominate the image.

The stretched image was loaded into *Photoshop* and the 'Dust & Scratches' filter (Filter > Noise > Dust & Scratches) applied with a pixel radius of 8 pixels. As seen in Figure 3.8, all the fainter stars disappeared, with the few remaining stars cloned out from adjacent areas of the image. This image was stored as 'Nebula'.

The original image was brought back and copied (Ctrl-A, Ctrl-C) and pasted (Ctrl-V) over the nebula image, producing two layers. These can be seen in the layers window which is opened by clicking on 'Window' in the set of options across the top of the screen and ticking the 'Layers' box. These were then flattened (Layer > Flatten Image) using the 'Difference' blending mode to give an image containing just the stars. To soften and slightly enhance the brighter stars, they were selected using the colour range tool (Select > Color Range[1]) by placing it over one of the brighter stars. The selection was increased in size by 4 pixels (Select > Modify > Expand), then the 'Gaussian Blur' filter (Filter > Blur > Gaussian Blur) was applied with a radius of

[1] American spelling.

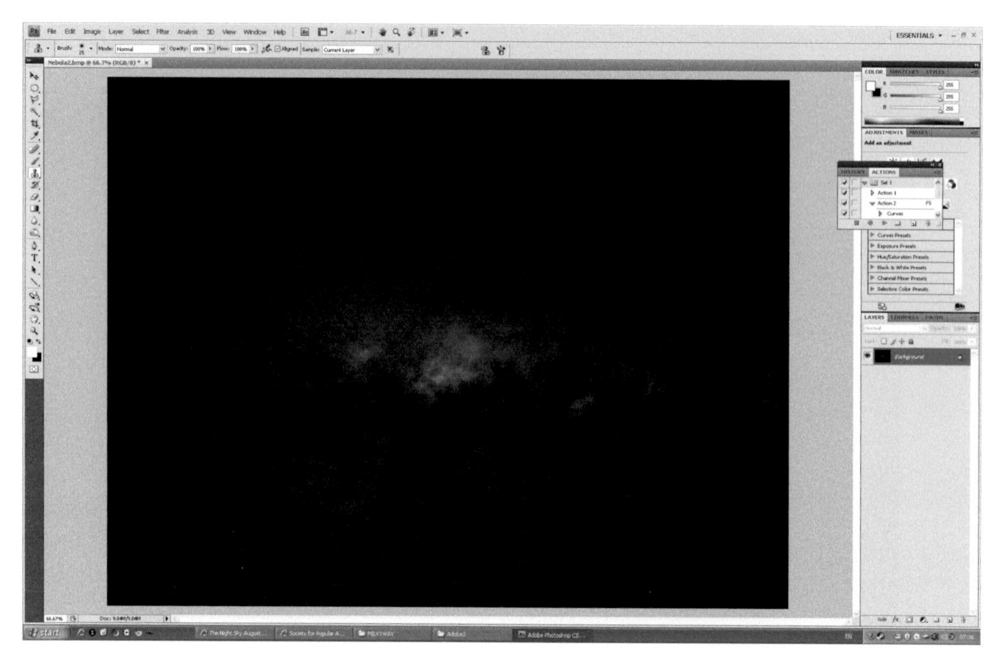

Figure 3.8 The 'Dust & Scratches' filter in *Adobe Photoshop* has removed the stars, leaving just the 'nebula' image

2 pixels. This reduces their brightness and so this was restored using the 'Brightness/Contrast' command (Image > Adjustments > Brightness/Contrast). The result was saved as 'Stars' and shown in Figure 3.9.

The 'Nebula' image was brought back and local contrast enhancement applied using the 'Unsharp Mask' filter (Filter > Sharpen > Unsharp Mask) with a large radius and small amount (say 250 pixels and 20 per cent). In the 'Levels' control window (Image > Adjustments > Levels) its contrast was increased by bringing the black level slider a little to the right and the mid tones slider a little to the left to give the result shown in Figure 3.10.

Finally the 'Stars' image was brought back and copied (Ctrl-A, Ctrl-C) and pasted (Ctrl-V) over the enhanced nebula image giving two layers. These were then flattened using the 'Screen' blending mode to add back the stars to the image. I discovered the use of the Screen blending mode some years ago when I was developing this technique by trial and error. The great thing is that, by using the 'Opacity' slider (set to 85 per cent for this image), the brightness and hence impact of the stars can be controlled so that they do not take away from the nebular regions at the heart of the Milky Way which were the target of this imaging exercise and seen in Figure 3.11.

Figure 3.9 Differencing the nebula image with the *IRIS* stretched image leaves just the stars, a little enhanced by using a 'Gaussian Blur' on the brighter stars

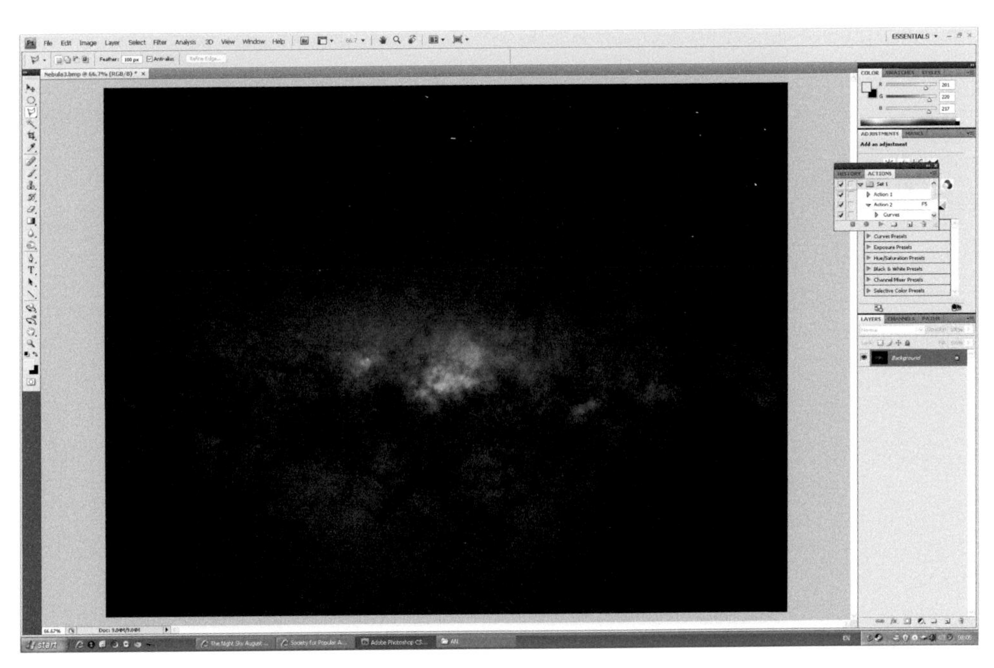

Figure 3.10 The nebula image is enhanced using the 'Unsharp Mask' filter and 'Levels' command

Figure 3.11 An image of the heart of the Milky Way taken from a dark site in New Zealand with a 40 mm equivalent, f/1.7 prime lens stopped down to f/2.8

4
Imaging the Moon with a Compact Camera or Smartphone

I suspect that rather few who read this book will use a compact camera to image the Moon but, as will be seen, I have produced some of my best lunar images using a 7 megapixel compact camera. Importantly though, this chapter will describe some image processing techniques that can equally be used when lunar imaging with a DSLR or video camera (webcam), so it should not be bypassed. It also describes the use of a program to build up a high resolution image from a number of segments (called panes) and which can also be applied to solar imaging.

The method for using a compact camera – or even a smartphone – is very simple. The basic idea, called 'eyepiece projection', uses a telescope with a low power eyepiece, focused on the Moon. The camera is held, either by hand or by using an adapter to mount the camera, so that its lens is aligned along the optical axis of the eyepiece, and an image taken. As the exposure will be very short, the technique can be used with undriven mounts such as Dobsonians.

Using a Smartphone

Smartphone cameras can find it very difficult not to overexpose a very bright object such as the Moon. This problem can be solved by using a camera 'App' to provide manual exposure control such as 'Camera+' for iPhones or 'Camera FV-5' for Android cameras such as the Samsung Galaxy series. Telescope adapters are available to mount them on the telescope eyepiece and so they can be used just as described below for compact cameras. However, most towns have shops, such as 'Cash Converters' or 'CEX' in the UK, which sell pre-used camera equipment and small compact cameras can be obtained at very low cost; these would be a better choice.

Using a Small Compact Camera

These can be used to give really good results. Firstly they weigh very little, so can be mounted on small telescopes and secondly, by use of their zoom, the image scale

Figure 4.1 Canon G7 compact camera mounted on a Baader 6030 Microstage Adapter and a monochrome image of the third quarter Moon taken with it

can be adjusted to nicely frame the Moon. When imaging the 10-day old Moon on 30 November 2014 to illustrate this chapter, I decided to use the simplest possible equipment: an 80 mm, f/6.8 refractor mounted on a very sturdy tripod, a 25 mm Plössl eyepiece, as is usually provided with a beginner's telescope, and a Panasonic FX12, seven megapixel, compact camera with a three times zoom. I acquired and focused on the Moon which was moving slowly across the field of view. Holding up the camera to the eyepiece showed that, until the zoom was extended somewhat, the image was strongly vignetted (darkened towards the edge of the field of view). I adjusted the tripod head so that the Moon was just coming into view and then waited until it lay in the centre of the field and took a few images.

This is a somewhat fiddly, but feasible, exercise and the imaging is far easier if the telescope is mounted on a tracking alt/az or equatorial mount and the camera is supported by, for example, a Baader 6030 Microstage Digital Camera Adapter, available for around £30, which will fit onto both 31.7 mm and 50.8 mm eyepieces, as shown in Figure 4.1, left. This clamps around the eyepiece and allows the lens of virtually any compact camera that has a tripod socket in its base to be aligned with the exit pupil of the eyepiece. The image of the third quarter Moon (Figure 4.1, right) taken with a Canon G7 camera has been used for a *Daily Telegraph* magazine cover. It is worthwhile using a two second delay between pressing the shutter and the taking of the image to allow the camera (and telescope) to settle. I always underexpose the images somewhat as a simple technique used to later enhance the image will brighten the highlights.

When the images are first looked at in a photo-editing program such as *Adobe Photoshop Elements, Adobe Photoshop* or the free program *GIMP*, they will tend to look

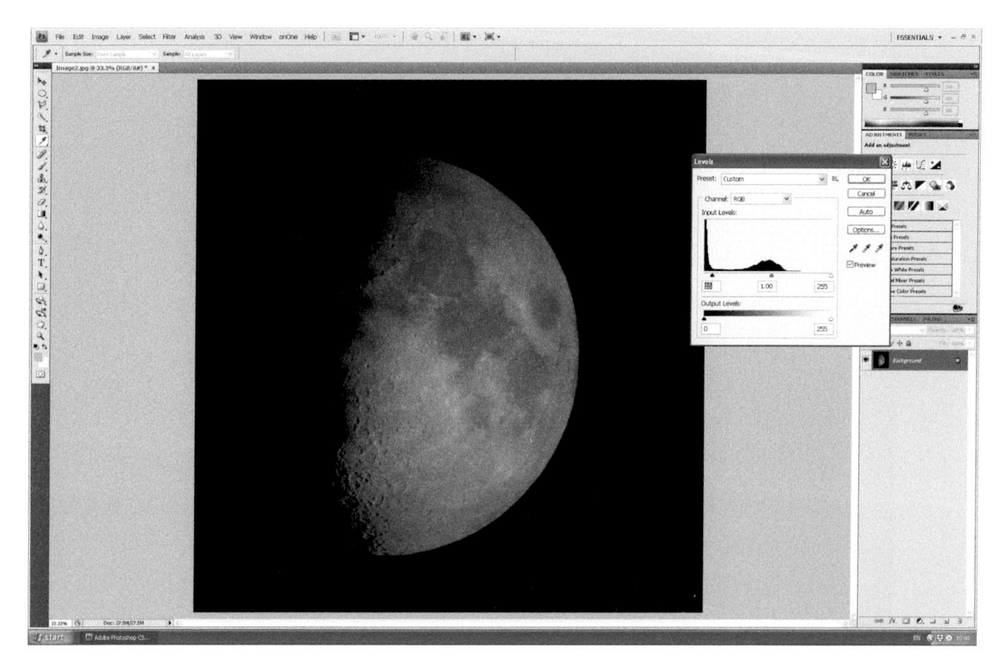

Figure 4.2 The use of the Levels command to increase the overall contrast of the image

a little 'flat' – the atmosphere scatters the moonlight and so reduces the overall contrast. Transparent nights will give the best results.

Increasing the Image Contrast

Lunar images can be greatly improved by a few simple techniques. The first is to increase both the overall and local contrast. First, using the 'Levels' command (Image > Adjustment > Levels), the 'black point' can be brought a little over to the right, as seen in Figure 4.2.

Then, given the use of the 'Unsharp Mask' filter (Filter > Sharpen > Unsharp Mask), the local contrast can be enhanced by setting the radius to a high value, say 200, and observing the result as the amount is increased. As seen in Figure 4.3, the result is quite impressive! This technique both darkens the *Mare* regions and brightens the highland regions – the latter being the reason that the captured lunar images should be somewhat underexposed. The histogram when applying the Levels command should peak at ~80 per cent, as is also seen in Figure 4.2. Lunar images often look better when converted to monochrome (Image > Mode > Greyscale).

Sharpening the Lunar Image

A little sharpening could then be applied within the same Unsharp Mask tool (Filter > Sharpen > Unsharp Mask) using a radius of a few pixels and adjusting the

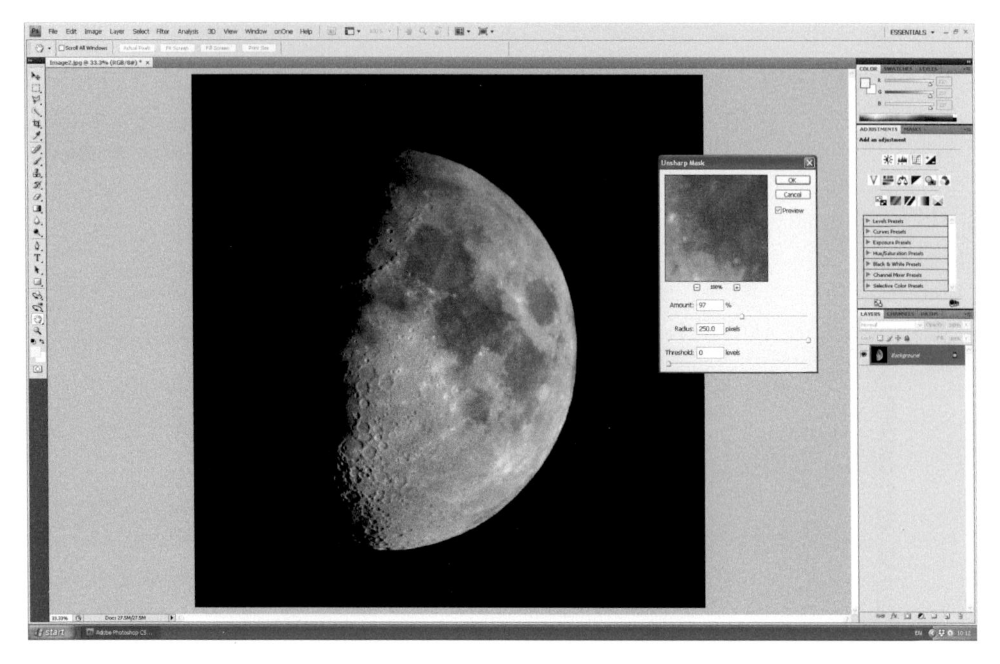

Figure 4.3 The result of using local contrast enhancement using the 'Unsharp Mask' filter

amount to suit. This is not the best method of sharpening and those with *Adobe Photoshop* or *GIMP* have other possibilities. The simplest to use in *Photoshop* is the 'Smart Sharpen' filter (Filter > Sharpen > Smart Sharpen). In the control panel that opens up, the 'Advanced' and 'More accurate' boxes should be checked, and in the 'Remove' menu, Gaussian Blur selected as seen in Figure 4.4. The radius (a few pixels) and amount are chosen to give an appropriate amount of sharpening.

Given the use of Layers and a High Pass filter in an image processing program, a further technique is to duplicate the layer and apply the 'High Pass' filter (Filter > Other > High Pass), with a radius of a few pixels, as seen in Figure 4.5. This gives a mid-grey image but the fine structure can be seen within it. The two layers are then flattened using the 'Overlay' blending mode. The amount of sharpening applied to the image can be finely adjusted by using the 'Opacity' slider in the levels box.

When using either of these methods, it is important not to over-sharpen the image!

Noise Reduction

When zooming into the *Mare* regions it may well be found that their floors show rather more noise than one might like. In *Photoshop*, the 'Despeckle' filter (Filter > Noise > Despeckle) can be used one or more times to smooth the noise. An alternative

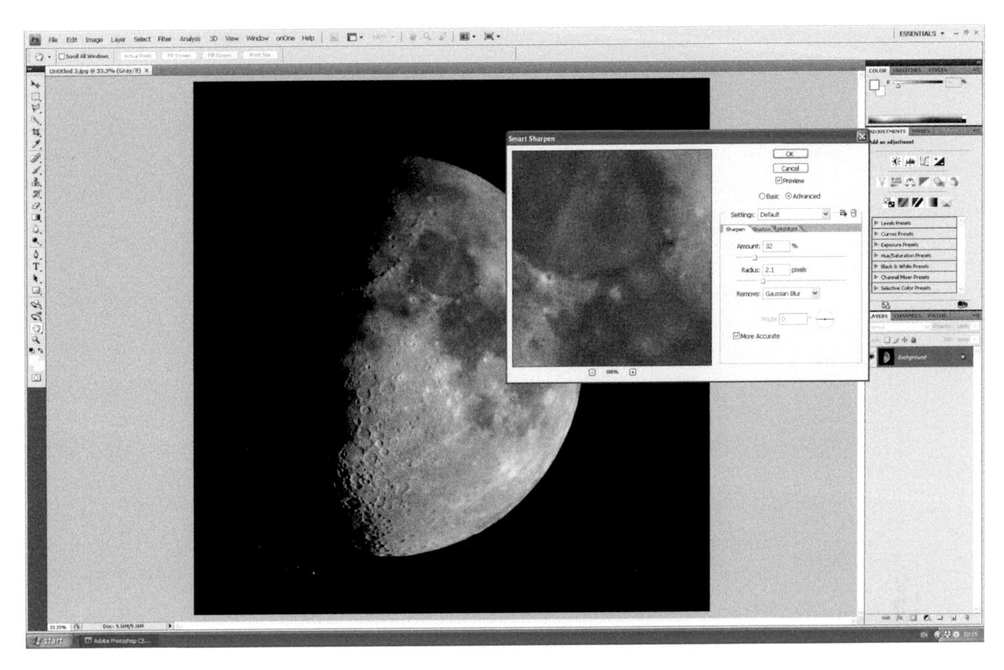

Figure 4.4 Using the 'Smart Sharpen' filter to increase the image sharpness

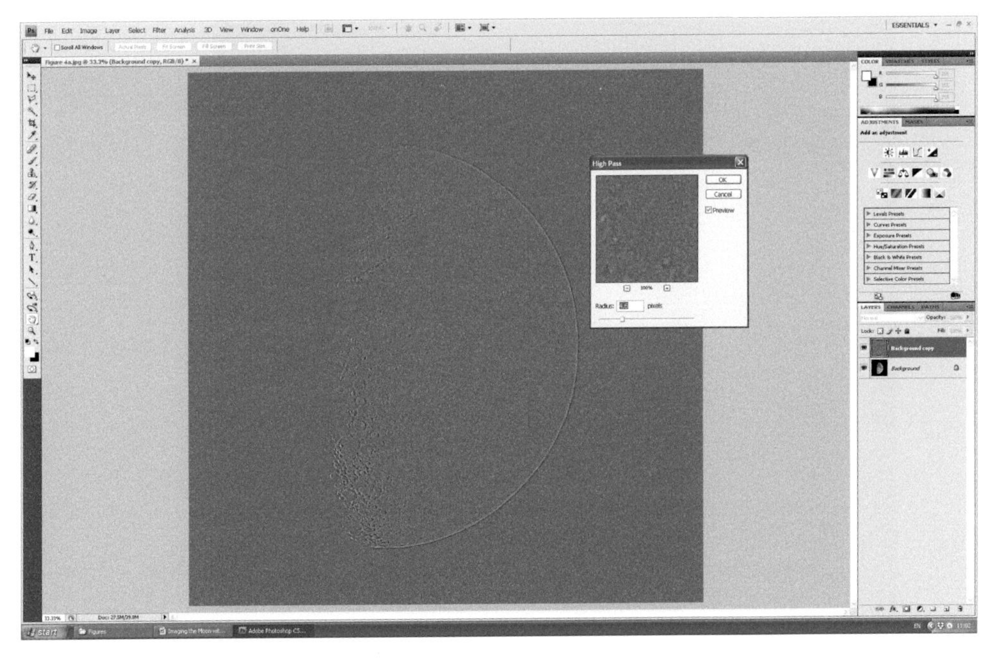

Figure 4.5 The use of the 'High Pass' filter to sharpen an image

Figure 4.6 The noise reduction program *Picture Cooler* can be used to smooth the lunar image

Figure 4.7 An image of the Moon taken with a hand held Panasonic FX12 compact camera and 80 mm refractor

is to use a noise reduction program such as *Picture Cooler*, which can be downloaded (for PCs only) from http://denoiser.shorturl.com/. This is widely regarded as one of the best noise reduction programs available. Figure 4.6 shows a lunar segment before and after the use of *Picture Cooler*: the top half as imaged and the lower half having had the noise reduction applied. Having applied a little further sharpening using the Smart Sharpen filter, Figure 4.7 is the final image – not too bad having used a hand held compact camera.

Making a High Resolution Composite Image

On the evening of 3 December 2014, the sky was transparent and the seeing was exceptionally good. This made it worthwhile to attempt a high resolution image of the 13-day-old Moon. Only one additional piece of equipment was required – a 2× Barlow lens. Placing this prior to the eyepiece doubled the effective focal length of the 80 mm telescope and thus doubled the image scale. As a result, when a mid-range zoom was used on the camera the lunar image was greater than the field of view. This meant that the camera was imaging a smaller area of the Moon, increasing the image scale on the sensor and so had the potential of increasing the effective resolution of the image.

To image the whole Moon, several images had to be taken to encompass the whole surface. These should have plenty of overlap and I actually took images of eight lunar segments to be sure to completely cover it. A tracking mount with a hand operated slew control is a great help in acquiring these. Each lunar segment was imaged five times and the sharpest of each selected to include in a mosaic covering the whole Moon.

I then opened up a superb, free, piece of software called *Microsoft ICE* (Image Composite Editor) which can be downloaded from http://research.microsoft.com/en-us/um/redmond/projects/ice/. The eight lunar segments were simply selected and dropped into its workspace. Within a few seconds *ICE* had produced an image of the complete lunar disc derived from six of the segments. (I had taken a couple more than needed.) It seems capable of handling any distortions and copes with slight changes of brightness from one segment to another. When I first used this program to produce a lunar mosaic I was amazed!

The image, as seen in Figure 4.8, was saved and the canvas size increased in *Adobe Photoshop* before being cropped to give a rectangular black background to frame the Moon. The same four enhancing steps were used as described above: black level adjustment, local contrast enhancement, slight sharpening and a little noise reduction. This image, Figure 4.9, is as good as any I have taken with far more sophisticated techniques and certainly proved to me that a small telescope and simple camera can give really good lunar imaging results.

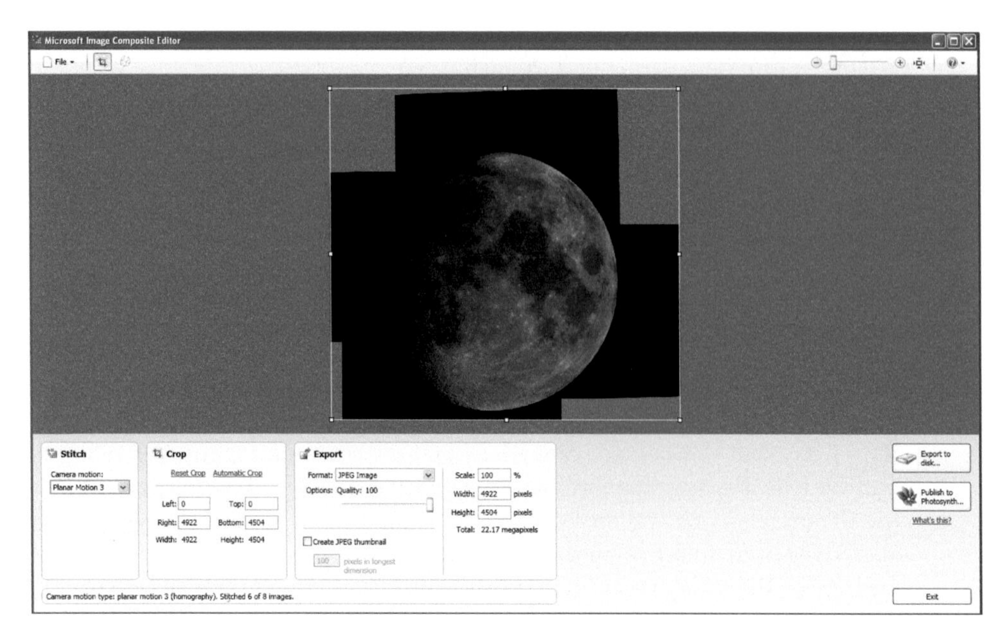

Figure 4.8 The output screen of *Microsoft ICE* having stitched 6 monochrome lunar segments

Figure 4.9 An image of the 13-day old Moon taken with a Panasonic FX12 compact camera

5
Imaging the Moon with a DSLR

In this chapter, I would like to describe how one can use a DSLR and telescope to image the Moon – the ideal first target for such a system. The DSLR is attached to a telescope using a T-mount comprising a 2 inch barrel to fit into the focuser and a bayonet which is specific to the make of DSLR, as seen in Figure 5.1. (Most telescopes now use focusers with a 2 inch barrel size, but it is possible to purchase T-mount barrels that will fit within a 1.25 inch focuser. The problem that can then arise is vignetting (the outer parts of an image will be darkened), but this is usually only a problem when imaging the Moon.) When using a refractor I have usually found that a 2 inch extension barrel, also shown in the figure, is required to bring the camera to focus.

The resolution of the image to be captured by the camera is dependent on two factors: the resolution of the telescope as determined by its aperture (~1 arc second for a 150 mm aperture telescope) and the 'seeing' on the night in question due to turbulence in the upper atmosphere. On a really good night, the atmosphere might limit the resolution to 1–2 arc seconds, while on a poor one this could be 3–5 arc seconds. The 'sampling theorem' states that to extract all the information in an image it needs to be sampled with a resolution half that of the resolution determined by the two factors described above – unlikely to be better than 2 arc seconds. The Moon has an angular diameter of 1800 arc seconds. The 'seeing' and telescope resolution will mean that there are, at best, 900 individual points across the lunar surface so, to double this and extract all the image quality possible, our camera needs to have 900 × 2 or 1800 pixels across the Moon's diameter. Modern DSLRs can easily accomplish this even if the Moon's image does not fully cover the field of view.

The lunar image on the camera sensor can be quite small if a short focal length telescope is used, thus possibly limiting the quality of the image. If this is so, a Barlow lens can be used to give a larger image. These typically give a magnification of ×2. It could well be that this will then give an image of the Moon that is too large to fit on the sensor. However, it is sometimes possible to reduce the magnification to, say, ×1.5 so that the lunar image will fit on the sensor. The magnification of a Barlow lens depends on its separation from the camera sensor and reducing this lessens the magnification. The lens element of a 2 inch Barlow can often be unscrewed and then screwed into the barrel of the T-mount, giving a magnification of ×1.5 rather than ×2.

Figure 5.1 Front – the bayonet and 2 inch barrel to attach the DSLR into a 2 inch focuser; back right – a 2 inch barrel extender that may be required; left – a 2 inch Barlow

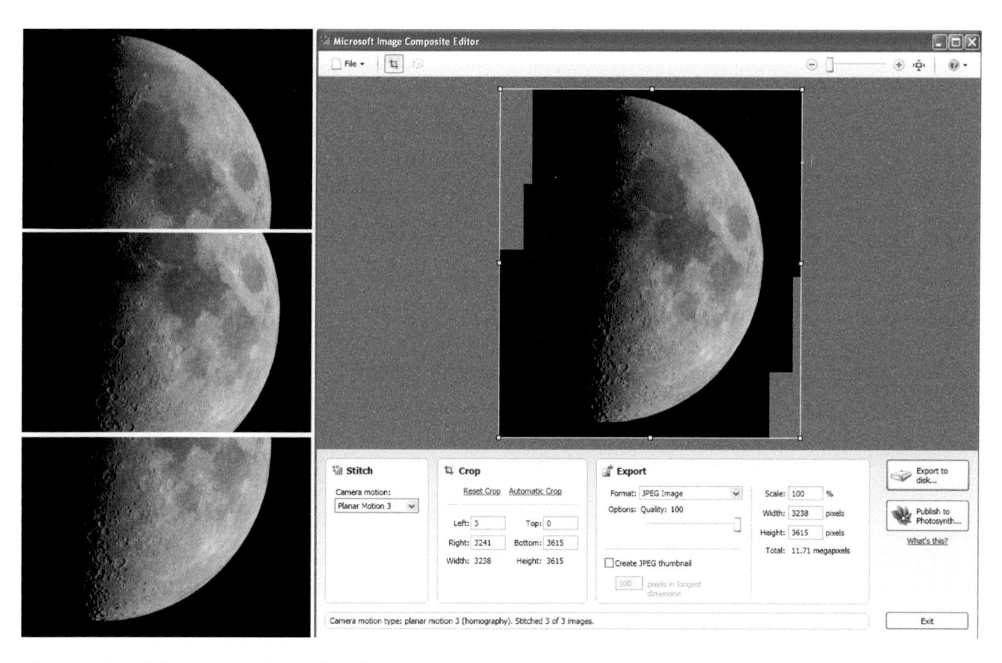

Figure 5.2 Three overlapping lunar segments with the screenshot of the *Microsoft ICE* program having combined them into a single image

Figure 5.3 The enhanced composite image

There is a further approach to overcome this problem. One can, for example, take three images (say) of the top, middle and bottom sections of the Moon (ensuring some overlap), as shown in Figure 5.2 (left). These are then merged into one using *Microsoft ICE* (described in Chapter 4), as seen in Figure 5.2 (right), to give the result shown in Figure 5.3.

Unless the telescope is equipped with a field flattener (to be discussed in Chapter 8), the outer parts of a full lunar disc may not be acceptably sharp. The solution then is to take several offset images so each part of the lunar disc lies in the central part of the field. These images are then cropped so that only the high quality part of each image is left. There has, of course, to be some overlap between these cropped images so that *ICE* can merge them into a single full disc image. For final processing of the lunar images, they can be imported into *Adobe Photoshop* and, as was described in detail in Chapter 4, some local contrast enhancement added by using the 'Unsharp

Mask' filter with a large radius and by a small amount (set by trial and error). As this procedure tends to brighten the highlights, it is best if the initial images are slightly underexposed. It can also be useful to use the 'Levels' function to increase the contrast a little (bring the 'black point' (left hand pointer) somewhat over to the right). Finally a little sharpening can be applied.

Sharpening Images in *Photoshop*

There are several ways of sharpening an image in *Photoshop* CS2 and above, two of which ('Smart Sharpening' and 'High Pass' sharpening) were discussed in Chapter 4.

Some professionals use a somewhat mystical process which, somewhat amazingly, uses a 'Blur' function to apply the sharpening! There are several steps in the process which can best be made into an Action as described below. I hope these steps will enable you to apply the effect.

1) Load the image and, should you wish to, initiate a new Action (as described below) to make the sequence into an Action so it can be easily replicated.
2) Make two copies: (Layer > Duplicate Layer), (Layer > Duplicate Layer) or (Ctrl-J, Ctrl-J).
3) Shift-click on the first copy so that also becomes blue and then click on the little menu box at the extreme top right of the Layers panel.
4) In the drop-down menu which results, click on the line 'New Group from Layers'.
5) In the window that appears, change the blending mode from 'Pass Through' to 'Overlay' and then click on OK.
6) The two copies are compacted into 'Group 1'.
7) Click on the arrow to the right of the eye in the group to open the two layers within the group and click on the upper layer of the two – it becomes blue.
8) Change the blending mode to 'Vivid Light' – the image looks disastrous.
9) Invert the layer using (Image > Adjustments > Invert) or (Ctrl-I). The image will now look normal again.
10) At this point stop recording an action if one is being made.
11) Now, the clever bit: open the 'Surface Blur' filter (Filter > Blur > Surface Blur) and adjust the radius and amount as if were was using the Smart Sharpen filter until an appropriate amount of sharpening results.
12) Finally flatten the layers (Layer > Flatten Image) to give the end result.

This looks pretty complicated but is not too difficult to do, particularly if most of the steps are made into an Action (for details of how to do this, see Making a *Photoshop* Action in Chapter 2). Quite how it works is not totally obvious, but the Surface Blur filter does treat areas of sharp detail differently to smooth areas, smoothing the latter but leaving the former untouched. The method does appear to lift up the darker regions of an image somewhat. I do think that it is worth a try.

Sharpening by Deconvolution

There is an alternative approach to sharpening – and even increasing the effective resolution – which attempts to estimate how the lunar surface (or other astronomical object) would have appeared without the resolution limits of the telescope used or the effects of the atmospheric turbulence. It works very well on lunar images. The captured image is a 'convolution' of the true image and what is termed a 'point spread function' that combines all the effects that would blur the image. As the seeing will probably dominate, this function is likely to be close to a Gaussian form. If this is known or estimated, then one can 'deconvolve' the captured image in an iterative process to remove, at least to some extent, the blurring and so produce an image with higher resolution. The process was extensively used to improve the Hubble Space Telescope images when it suffered from an incorrectly shaped mirror. There are a number of algorithms that can be used and I have tended to use the Lucy–Richardson algorithm. The only downside to the deconvolution process is that it can take some time to implement and so a reasonably fast computer is advised.

I have been using the set of deconvolution filters that are provided in the software suite that can be purchased from 'Astra Image' (www.astraimage.com) for under £10. They can either be accessed as a 'stand-alone' package or as a set of 'plug-ins' for *Adobe Photoshop*. As the point spread function will not be known, *Astra Image* allows one to be set (often a Gaussian function) and helpfully provides a preview screen that allows the effects of changing the control parameters to be quickly seen. A trial version that watermarks the results can be downloaded for free and is very well worth a try.

Taking Lunar Images

If taking an image manually, it is best to apply a two second delay between pressing the shutter button and the taking of the image to allow any vibrations to die down and perhaps also to include a one second delay between the mirror rising and the exposure. If, as may well be possible, 'Live View' can be used, the mirror is kept raised and, usually, the Live View image can be zoomed into, which is a great help in achieving focus on the lunar disc. A remote release, such as provided by the intervalometer described in Chapter 1, is also useful.

Short exposures with a high ISO will help ameliorate the effects of seeing, but can tend to be noisy. If this is a problem, the 'Despeckle' filter in *Photoshop* or a noise reduction program such as *Picture Cooler* can be used to smooth the image. The Moon is not colourless and, if the saturation is increased, a quite colourful image can result, as shown in Figure 5.4. This actually tells us something about the geology of the lunar surface as, while *Mare Serenitatis* appears reddish-brown due to iron compounds in its basalt, *Mare Tranquillitatis* appears somewhat blue caused by richer titanium-bearing minerals. Often, however, lunar images tend to look better when converted to monochrome.

Figure 5.4 A lunar image with increased saturation showing the colour variations across the surface

Another good lunar imaging target is to image the 'earthshine' that is sometimes visible hanging between the limb of a thin crescent Moon. The image of Figure 5.5 was taken from a dark sky site using a Takahashi FS102 refractor having superb overall contrast (partly because of the use of fluorite doublet) which prevented the light from the bright limb washing out the far fainter earthshine. One exposure was made to capture the bright limb and a longer one to capture the earthshine. The two were then combined in *Adobe Photoshop*.

Figure 5.6 is a single lunar image taken using a Nikon D7000 DSLR mounted on a 127 mm apochromat refractor and Teleskop Service 2 inch Field Flattener. The atmospheric seeing was superb and the Moon was high in the sky, so a very high quality image resulted.

Figure 5.5 An image of earthshine taken with a Takahashi FS102 and Nikon D7000 camera from a dark sky site

Figure 5.6 An image of the Moon taken on a night of superb seeing with a CFF Telescopes 127 mm apochromat refractor and Nikon D7000 camera

6
Imaging the Pleiades Cluster with a DSLR and Small Refractor

The Pleiades, also known as the Seven Sisters or Messier 45, is a very good first 'deep sky' imaging project for a newcomer to astrophotography. As the Pleiades has quite a large angular size, small refractors such as those with an 80–100 mm aperture will be ideal.

Choosing a Suitable Focal Length Telescope for a Celestial Object

Of course, if one only has one telescope, one may not have any choice, but it is quite likely that, over time, an astrophotographer will acquire several telescopes of differing focal lengths, and hence varied fields of view when used with a DSLR or CCD camera. Large angular sized objects such as the Pleiades cluster, M45, the North America Nebula, NGC 7000 and the Andromeda Galaxy, M31, will be best imaged with a short focal length refractor. Small angular sized objects such as planetary nebulae or globular clusters, examples being the Ring Nebula, M57, the Dumbbell Nebula M27, and the globular cluster in Hercules, M13, really need to be imaged with a long focal length telescope such as an 8 inch Schmidt–Cassegrain. Having one of each would cover most astroimaging challenges. Adding a telescope of intermediate focal length, such as a 102, 120 or 127 mm refractor or possibly a 20 mm, f/4, Newtonian (used with a coma corrector) would allow virtually any object to nicely fit within the camera sensor.

Over the years, I have collected quite a number of telescopes, and those that will be used for the imaging examples within this book are: short focal length refractors of 66, 72 and 80 mm aperture; intermediate focal length refractors of 102 and 127 mm aperture; a 200 mm aperture, f/9, Vixen VC200L astrograph; and a Meade 8 inch Schmidt–Newtonian. The majority have been bought second hand in the UK from the 'UK AstroBuySell' website (http://www.astrobuysell.com/uk/), which allows users to sell and buy virtually everything required by an amateur astronomer – usually with a good saving on new prices.

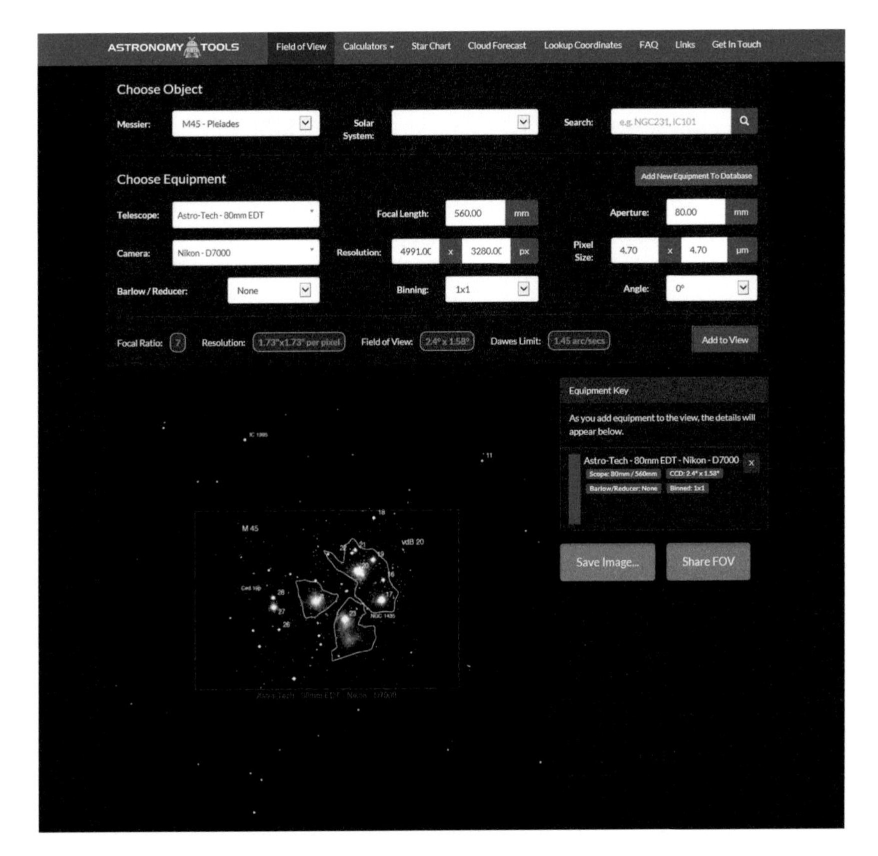

Figure 6.1 Having been given details of the telescope and camera to be used to image a specific object, the *Astronomy Tools* field of view calculator shows how it will be framed by the camera sensor

Field of View Calculator

An excellent piece of software to give the field of view of a very wide range of telescopes and cameras is provided within the *Astronomy Tools* software package: http://astronomy.tools/calculators/field_of_view/. First, the object to be imaged is selected, in this case, M45. The telescope that it is proposed to be used (or one with a similar focal length) and the camera that is to carry out the imaging are selected from two very extensive lists. Again, if the specific camera is not included, one is chosen with the same sized sensor – typically a Nikon or Canon APSC sensor (these have slightly different sizes, the Canon being smaller). Here an 80 mm aperture telescope, having a focal length of 560 mm was to be used allied to a Nikon D7000 camera with APSC sensor. When the 'Add to View' tab is clicked on, a chart is displayed showing the chosen object in the centre of the field of view of the telescope–camera combination, as seen in Figure 6.1.

Using an Alt/Az Tracking Mount

A tracking mount is necessary, but this could be either an alt/az (altitude/azimuth) or an equatorial. One thing to be aware of is that, when using an alt/az mount, the sensor will rotate slightly with respect to the object being imaged during the sequence of exposures. This is termed 'frame rotation'. If one observes Orion as he rises, he will be seen to lie on his right hand side; when due south Orion will stand vertical; and as he sets, he will lie on his left hand side. So, relative to the sensor of a camera mounted on an alt/az mount, Orion will have rotated through 180 degrees. Of course, one will only image an object for a number of minutes, perhaps up to an hour, and the effect is not so serious. The effect of the frame rotation is removed by the stacking software, which can rotate the individual frames so that they are aligned before they are stacked. However, the result will show that areas in the corners of the frame will not have had as much exposure as that in the centre and the resulting image will need to be cropped somewhat. Rather annoyingly, frame rotation is least when an object is rising in the east or setting in the west, but has the greatest effect when it is due south. This is, of course, exactly when one would like to image an object, as it is then highest in the sky and so the image will be less troubled by turbulence in the atmosphere.

Alt/az mounts are discussed in Appendix B. The mount that was used for this imaging project was the iOptron MiniTower, which is exceedingly quick to set up when making an expedition to a dark sky site (in this case in Shropshire) and can handle refractors up to 120 mm in aperture or a small Maksutov or Schmidt–Cassegrain such as the Celestron 6 inch.

Using an Equatorial Mount

The two advantages of using an equatorial mount, discussed in detail in Appendix B, are that they do not suffer from frame rotation and that only one drive motor, running at a constant rate, is required to track an object across the sky. This is in contrast to an alt/az mount, where two motors running at variable speeds are required. The tracking (how well the mount follows the object being imaged) is thus likely to be more precise. The disadvantage is that they do require some setting up to align them on the North Celestial Pole and so are perfect when used in a fixed location but perhaps not so suitable for a brief imaging excursion to a dark sky site. (The use of a new product, the QHY PoleMaster also discussed in Appendix B, can accurately align an equatorial mount in just a few minutes so this statement may not now be so true.)

Exposure Lengths

If images are to be taken with an unguided mount, an exposure length should be chosen such that star trailing does not become apparent. Except in a very few cases this will determine the maximum exposure length. Exposure times of 20–40 seconds will normally be short enough to minimise tracking errors, and this will depend on how well an equatorial mount is aligned on the North Celestial Pole and the 'periodic error' produced by its worm gear or belt drive. If the mount is being autoguided (as discussed in Chapter 9 and Appendix D), the length of exposures will depend on other factors. One important consideration, particularly when imaging the Orion nebula, covered in the next chapter, is that important parts of the image must not be overexposed. A second consideration to bear in mind is that of the camera sensor's 'dark current'. This is composed of thermal electrons that fill up the pixel wells in the sensor even when no object is being imaged. The word 'thermal' is important as it depends dramatically on the sensor temperature and typically doubles for each 6 degree Celsius increase in temperature. This subject is considered in detail in Chapter 10 but the dark current of an uncooled sensor (which will stabilise at about 12° Celsius above ambient temperature) will be greatest on warm summer nights and least on cold winter nights. Shorter exposures will thus tend to be needed in summer. Chapter 10 will also show how a DSLR can be cooled at very low cost to reduce the dark current. Considerations as to how to avoid camera shake, use of raw and/or JPEG file capture and the choice of ISO to use were discussed in Chapter 2.

Imaging the Pleiades Cluster

To enable the faint blue nebulosity surrounding the brighter stars of the cluster to be brought out, a dark, transparent, sky with little light pollution will be required along with a total exposure of at least 12 minutes. With an unguided mount this might be made up of twenty-four, 30-second exposures. Chapter 12 will discuss the use of light pollution filters in light polluted skies, with the Astronomik CLS filter being one of the best. A blue filter could also be used to minimise its effects and will work as the light pollution is largely in the orange part of the spectrum and the nebulosity is blue. However, the situation in the UK is changing as orange sodium street lights are being replaced by LED lights. Though the scattered light from these is less (which is a good thing), they have a far wider spectrum of light and often have a peak in the blue part of the spectrum which is not rejected by the Astronomik filter. This may become an increasing problem for astroimagers in urban areas.

Combining the Light Frames in *Deep Sky Stacker*

The program used to combine the images is called *Deep Sky Stacker*. This can be downloaded for free as described in Chapter 2. *Deep Sky Stacker* will accept the majority of raw formats, but should it not accept those from your camera, a further free program called *Adobe DNG Converter* can be used to convert them into the '.dng' form which can be read by *Deep Sky Stacker*. One nice feature of *Deep Sky Stacker* is that it will attempt to remove any 'hot pixels' from the image, so partly removing the requirement for long-exposure noise reduction in-camera.

The raw or JPEG images are selected and loaded into *Deep Sky Stacker* by clicking on 'Open picture files'. Then, 'Check all' is clicked upon followed by 'Stack checked pictures' as seen in Figure 6.2.

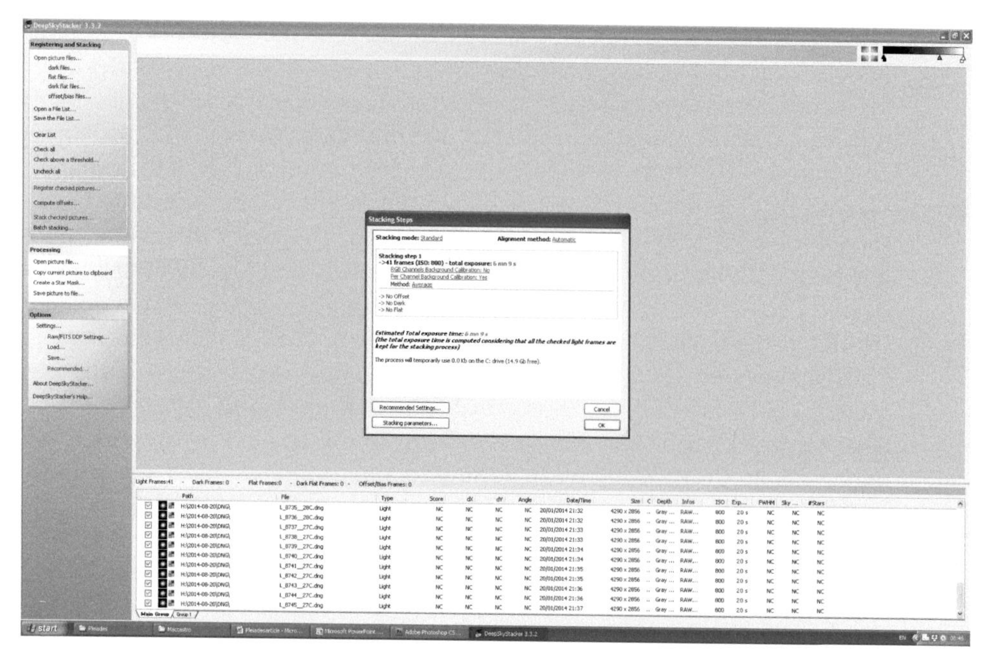

Figure 6.2 The raw or JPEG images are loaded into *Deep Sky Stacker*

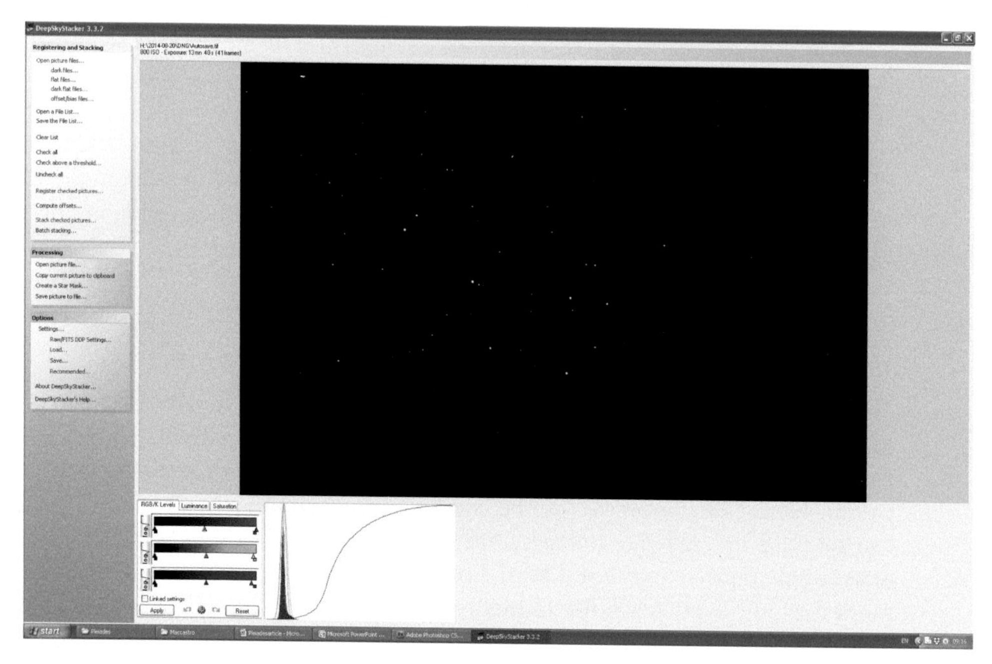

Figure 6.3 The final screen of *Deep Sky Stacker* as the stacked image is saved

Figure 6.4 The TIFF output from *Deep Sky Stacker* when loaded into *IRIS* produces a stretched image that can be saved as a JPEG file

Deep Sky Stacker analyses each image and aligns them before they are stacked to provide a deeper image which is then exported and saved as a 16-bit TIFF file (Figure 6.3). This often looks somewhat disappointing as fewer stars will be obvious than in a single JPEG image! However, they are not lost but, as the total dynamic range of the image is now far greater, they are just not so visible.

Herein lies a problem. Low cost imaging processing programs may import 16-bit files but cannot retain this bit depth when processing the images – which is vital if the fainter stars and nebulosity are to be brought out. However, a free program called *IRIS* can achieve just this. The 16-bit TIFF file is loaded into *IRIS* and it appears, as seen in Figure 6.4, that (somewhat surprisingly) the program will immediately apply a stretching function to the image so that the fainter stars and, hopefully, the nebulosity will be visible. This could be all that is needed and, if the image is then exported as a JPEG (but not TIFF) file, the resulting image can be imported into an image processing program for any final adjustments. If required, *IRIS* can provide more control over the stretching process by clicking on the 'Logarithm' command in the 'View' drop-down menu. Then, using the threshold sliders, the stretching can be controlled to optimise the image.

As the nebulosity is very faint, the stretching applied to make it visible will also tend to make the image noisy. *IRIS* can also help remove the noise by clicking on the 'Selective Gaussian filter' command in the 'Processing' drop-down menu. The two parameters 'Radius' and 'Threshold' can then be adjusted. I simply used the default

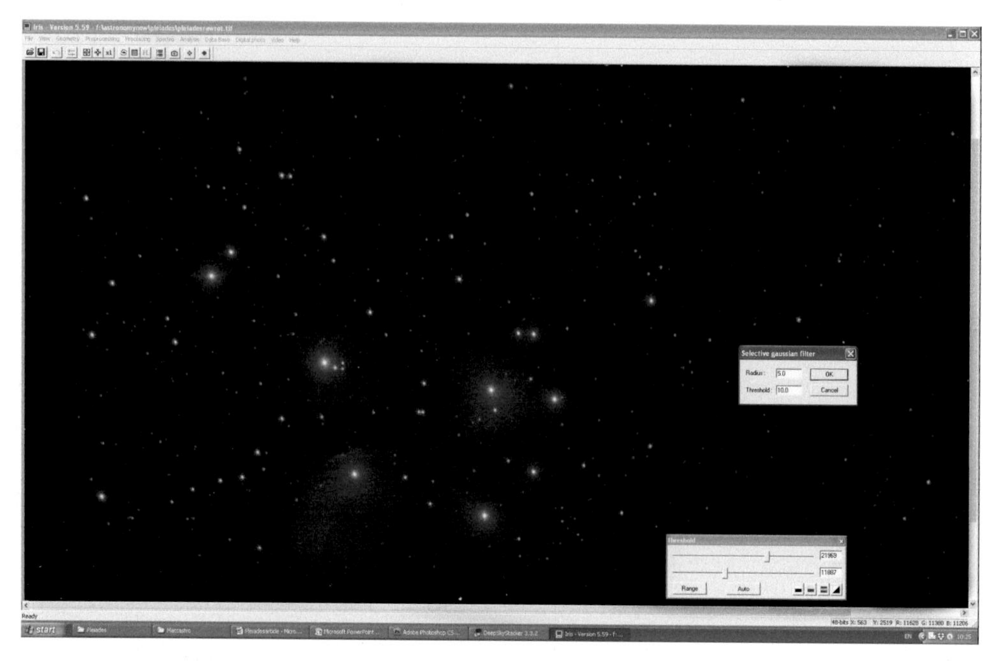

Figure 6.5 *IRIS* has stretched the image and applied noise reduction

values of 5 and 10 respectively in applying the noise reduction to achieve a final image. The image as seen in Figure 6.5 can then be saved as either a TIFF or JPEG file. The *IRIS* website has a good tutorial section to help.

Should you have *Adobe Photoshop*, then the stretching can be achieved by importing the 16-bit TIFF image and using the 'Levels' command. The centre slider is moved a little to the left to give a value of 1.20 and applied. Each time the process is repeated (it could be made an Action), the fainter parts of the image become more apparent. *IRIS* could then be used to reduce the noise in the resulting image as described above.

In the result given by the use of either program the background noise may well be too bright. In the Levels command of all image processing programs, the 'black point' controlled by the left slider in the levels histogram can be brought a little to the right to darken the background.

To produce the final image, it was cropped and the saturation increased slightly to bring out the nebulosity. When I took this image of the Pleiades from a relatively dark site in Shropshire, the clouds came in after taking only forty-one, 20-second exposures, giving a total exposure of just 13.6 minutes. Even with such a short exposure, the nebulosity shows up well. Streaks are seen within the nebulosity. These are real: caused by ferromagnetic dust particles acting like little magnets and aligning themselves along the magnetic field lines of our Galaxy. So this image, taken with a small telescope, tells us both that the dust in the interstellar medium must contain iron, and also that our Milky Way Galaxy has a magnetic field.

Figure 6.6 The Pleiades cluster taken with a Nikon D7000 DSLR mounted on an 80 mm ED refractor with 13.6 minutes total exposure

7
Imaging the Orion Nebula, M42, with a Modified Canon DSLR

The Orion Nebula (Messier 42) is one of the classic objects to attempt by a beginner to astroimaging but is, ironically, one of the hardest to image well. The reason is the great range in brightness between the central region that surrounds the stars of the Trapezium star cluster and the faint nebulosity that frames it and adds so much to the image. This chapter will also explain how a camera can be modified to capture more of the deep red hydrogen-alpha emission (from gas excited by the ultraviolet light from the Trapezium stars) that enhances the image.

Improving the Capture of H-alpha Emission by a DSLR Camera

DSLR sensors are sensitive to infrared (IR) light, which comes to a different focus to that of the visible spectrum and so would blur the image if allowed to fall on the sensor. To avoid this, all DSLRs have to incorporate a filter, placed in front of the sensor, which cuts off the infrared. But there is a second aspect to this filter due to the fact that our eyes are not that sensitive to red light, becoming less sensitive the longer the wavelength. As is shown in Figure 7.1, in order that the images taken by a DSLR match what would be seen by the human eye, the IR rejection filter has a gentle slope across the red part of the spectrum before reaching zero transmission in the IR band of the spectrum. The problem is that, as a result, the sensitivity of the sensor to hydrogen-alpha (H-alpha) emission is reduced to about 24 per cent.

Canon and Nikon Astrophotography Cameras

Both Canon and Nikon produce cameras adapted for astrophotography which use a sharper cut-off filter to allow greater sensitivity to the hydrogen-alpha emission. The Canon EOS 60Da, using an APSC sized senor, incorporates a modified low-pass filter having increased H-alpha sensitivity that is approximately three times higher than that of a normal DSLR camera. This 18 megapixel camera has a vari-angle LCD screen and can still be used for normal photography if an appropriate white balance is employed. At considerably higher cost, Nikon have the D810A camera which has a

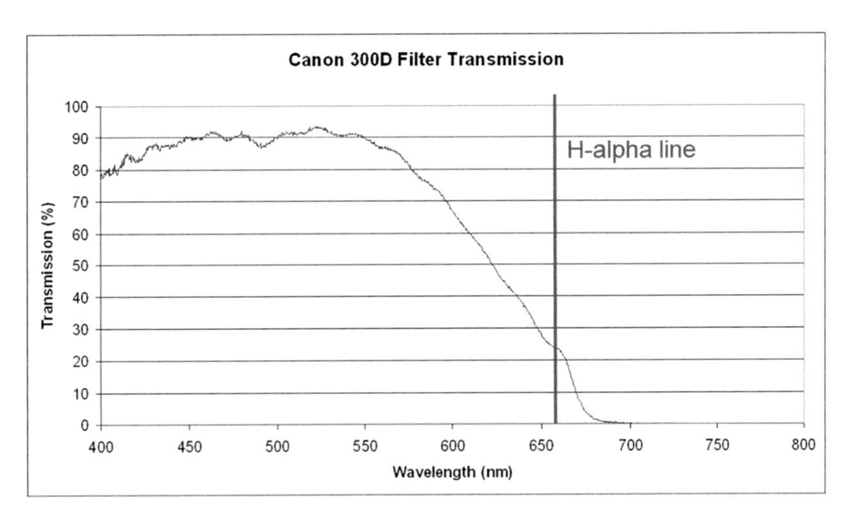

Figure 7.1 The spectral sensitivity of a typical DSLR, showing the lack of sensitivity at the hydrogen-alpha spectral line

36.3 megapixel full frame sensor with a redesigned infrared (IR) cut-off filter that is four times more sensitive to the H-alpha spectral line than the standard D810.

Modified Canon DSLR Cameras

Owing to the fact that the majority of astrophotographers have been using Canon DSLRs, a number of third-party companies will now modify standard Canon cameras to increase their sensitivity to H-alpha emission. This involves removing the standard IR filter and replacing it with a filter (ref. #245 9211) made by Baader Planetarium. This does, of course, nullify the Canon warranty. As shown in Figure 7.2, the Baader filter passes almost 100 per cent of the H-alpha emission and then cuts off very steeply into the infrared. Instructions to carry out the modification can be found on the Internet, but I would advise having it done professionally. One camera that has been widely modified is the Canon 1100D and modified examples can often be acquired at low cost from the UK AstroBuySell website. At the time of writing, a modified, 18 megapixel, Canon EOS 100D is probably the most cost effective camera to use for astroimaging and, as an example, Astronomiser in the UK can provide a new modified camera body with their own 1-year warranty. They will also modify most Canon cameras. A modified Canon 1100D was used to make the image described in this chapter, but using an unmodified DSLR will also give a very nice result – it is just that the H-alpha emission will not be so apparent.

Imaging the Orion Nebula, M42

In many images of the Orion nebula that are seen on the Internet, the central region is burnt out as the exposure required to bring out the nebulosity is far greater than that

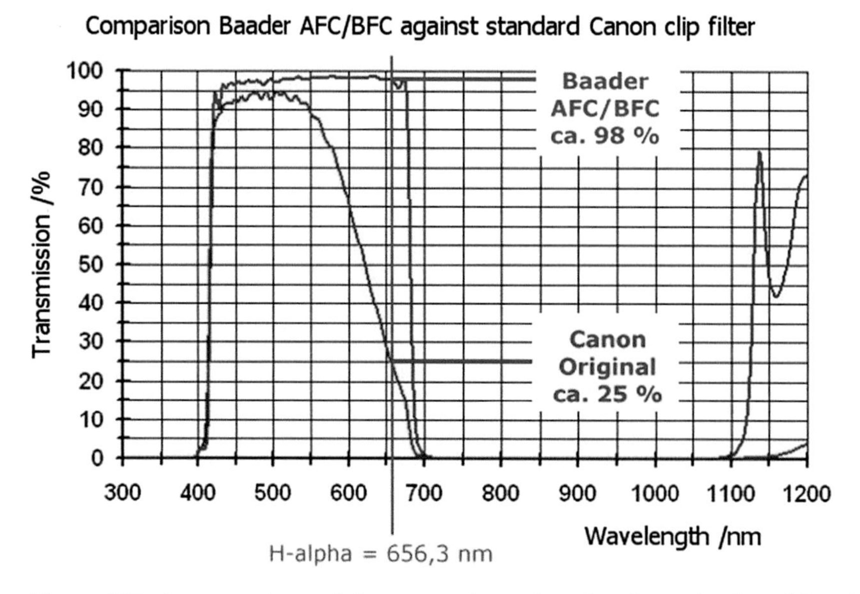

Figure 7.2 A comparison of the spectral passbands of standard and Baader IR filters

need the correctly expose the central region. There are two ways of imaging M42 to cope with the very wide range of brightness. One is to take a series of images of different exposures. In the shortest, perhaps only 12 seconds long, the central region will be correctly exposed, while longer exposures will capture the brightness range of the nearby nebulosity and fainter outer regions. To combine these together is a somewhat complex process using layers and masks in *Adobe Photoshop*, but there is an alternative approach. This is to take many very short exposures, stack them in the free program *Deep Sky Stacker*, as described in previous chapters, and then stretch them in *IRIS* or *Photoshop*. For this Orion Nebula image, a modified Canon EOS 1100D was used with a 102 mm, f/8 refractor using an unguided mount. Using an ISO of 800, thirty-seven, 9-second exposures (called light frames) were taken so that the central part of the image including the Trapezium Cluster was correctly exposed. The light frames were exported as raw files and loaded into *Deep Sky Stacker*. As the exposures were very short, there was no need to take 'dark' frames and, with an f/8 focal ratio, 'flat' frames were not required either. (Dark and flat frames are discussed fully in Appendix E.) The result is shown in Figure 7.3.

As usual with *Deep Sky Stacker*, only the very brightest parts of the image appear. This image was exported as a 16-bit TIFF file and imported into the free program *IRIS*. The initial, automatically applied, stretching of the image greatly overexposed the central region, but when in the 'View' drop-down menu the 'Logarithm' command was selected and the two range sliders were adjusted, a very presentable stretch was made, as shown in Figure 7.4. This was exported as a TIFF file to import into an image-processing program such as *Photoshop*. I trust that you can see that the technique of

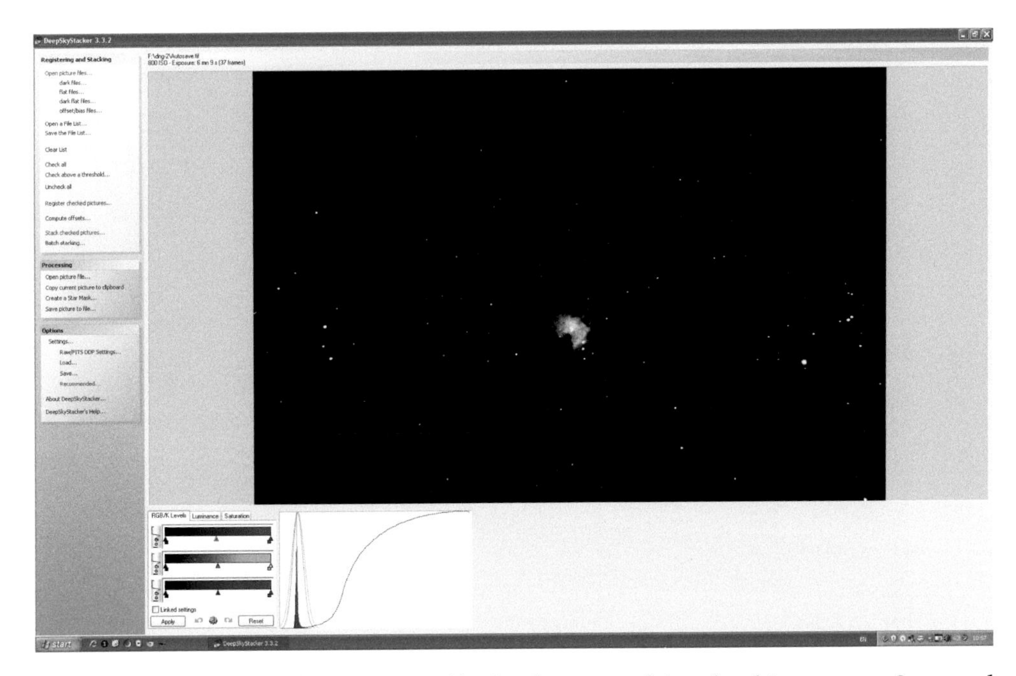

Figure 7.3 The result of using *Deep Sky Stacker* to combine the thirty-seven, 9-second exposures

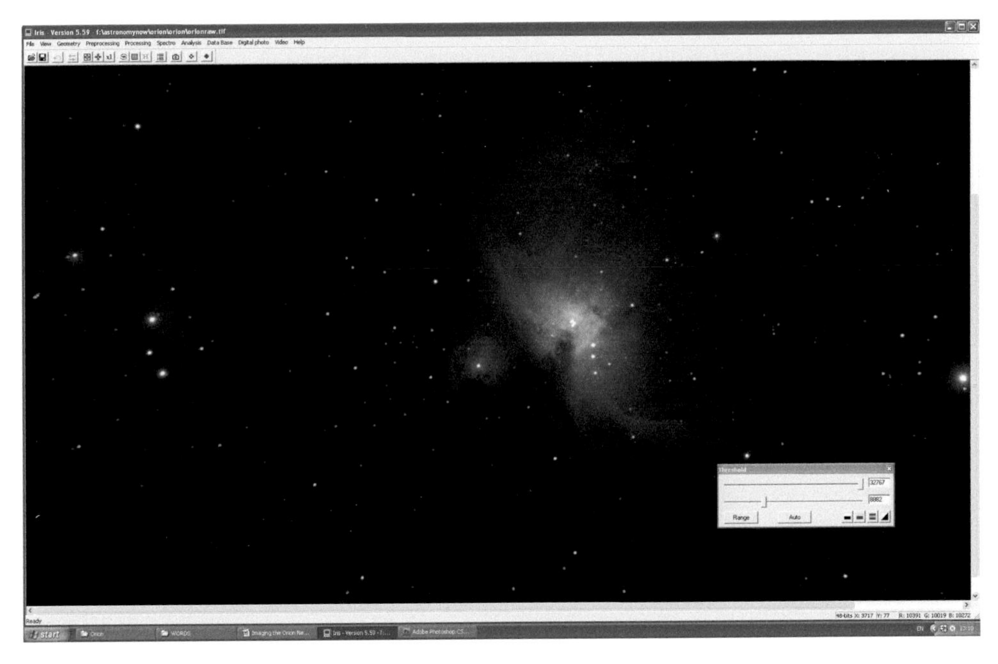

Figure 7.4 The result of using the *IRIS* program to 'stretch' the image and so bring out the faint nebulosity

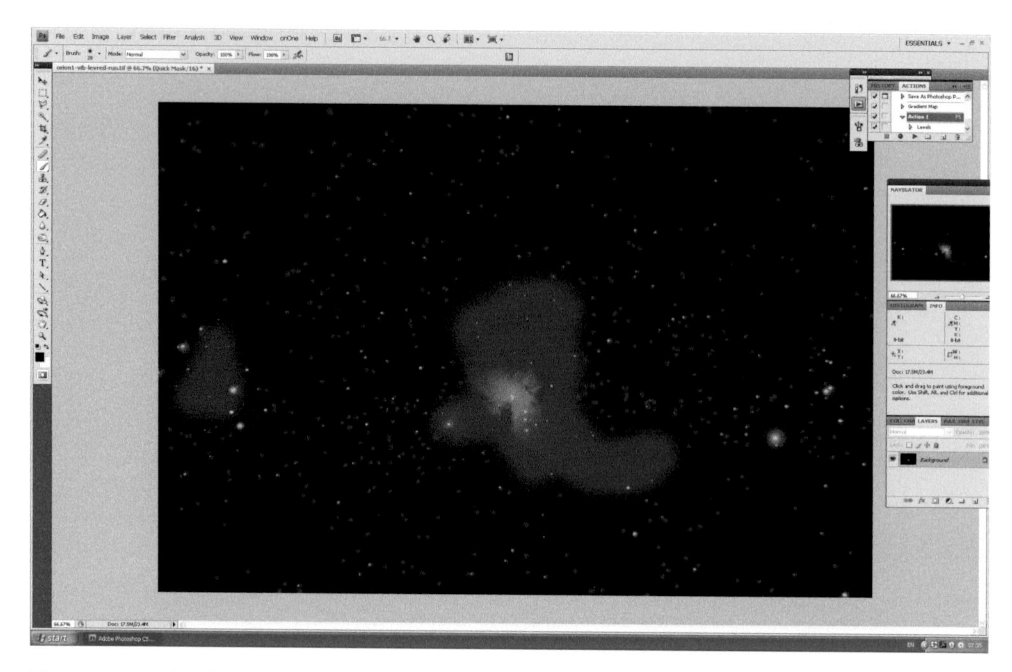

Figure 7.5 The bright nebula regions and stars were masked out before applying a Gaussian Blur to reduce the noise in the fainter parts of the image

'stretching' an image is fundamental to bringing out the faint nebulosity that adds so much to the final quality of many astronomical images.

Unless the total exposure is very long, the fainter parts of the nebula will be rather noisy. For this image a somewhat time consuming, but very effective, noise reduction method was applied in *Photoshop*. Using the 'Polygonal Lasso Tool', the central region of the nebula where the noise was not apparent was initially selected. A further selection was added around the region of the 'Running Man' Nebula with the shift key held down. This selection was inverted and the selection converted into a mask by clicking on the Mask (circle in square) icon at the bottom of the tool bar. The outer (noisy) parts of the nebula will be visible with the protected area masked in red. Then with a small diameter paintbrush, all the stars were masked, so covering the image with red spots as seen in Figure 7.5. (I did say that it was a bit time consuming!) The mask icon was clicked on again to show the 'dancing ants' surrounding each of the stars and the central region. The outer part of the nebula where the noise was apparent was zoomed into and a Gaussian Blur applied (Filter > Blur > Gaussian Blur), choosing an appropriate pixel range by trial and error (three pixels in this case), to smooth out the noise.

Two further enhancement techniques were used. The nebula and 'Running Man' regions were selected, the selection converted to a mask as before and the stars within them deselected using the 'Paintbrush' tool. (Deselecting the stars prevents them becoming more prominent.) Having converted back to the 'marching ants' display,

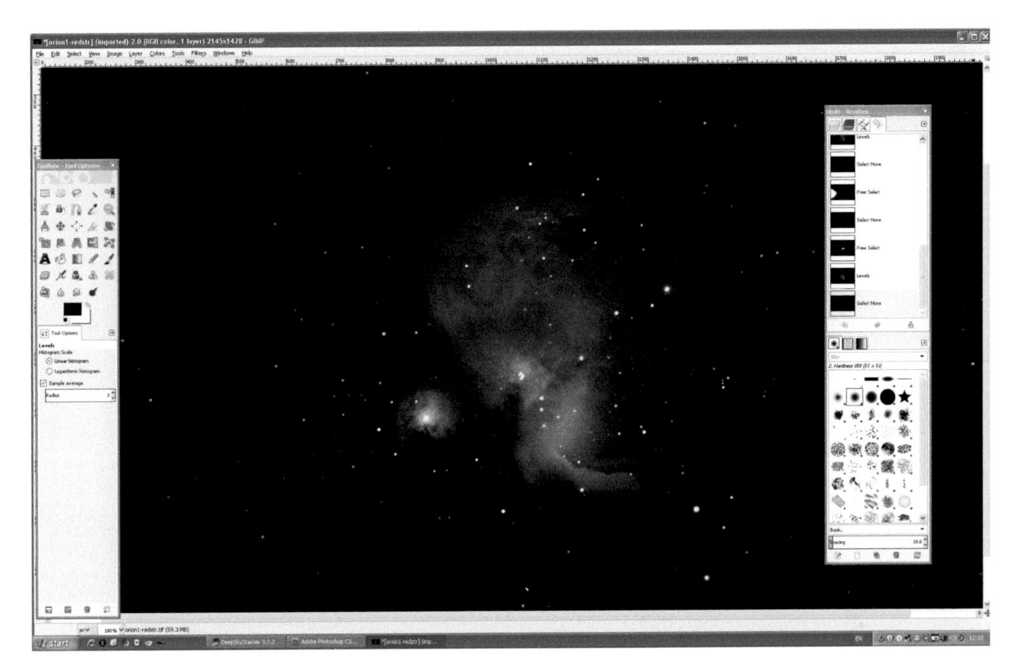

Figure 7.6 The result of using the free program *GIMP* to enhance the nebula image

'local contrast enhancement' was then applied using the 'Unsharp Mask' filter (Filter > Sharpen > Unsharp Mask) with a radius of ~200 pixels and an amount of ~40 per cent. (These are adjusted whilst observing the result.) With the same mask, the image saturation was increased (Image > Adjustments > Hue and Saturation) to heighten the colour. One pleasing thing about the processing of this imaging example is that this can be partly achieved using the free program, *IRIS*, that will work with 16-bit files to produce a very good stretched image. This can then be imported to a low-cost or free image processing program (such as *GIMP*) to apply the enhancements that provide an attractive final result. For example, Figure 7.6 is included to show the pleasing result, using *GIMP*, of applying some local contrast enhancement to the nebula region.

The main image (Figure 7.7) shows the Orion Nebula, M42, with the four stars of the Trapezium at its heart. To their left is the dark cloud known as the 'Fish's Mouth' which separates M42 from M43, another smaller nebula centred on the young irregular variable star Nu Orionis. Above them lies a blue reflection nebula known as 'the Running Man', which shows up well in the image.

Figure 7.7 The sword of Orion captured with an H-alpha modified Canon 1100D DSLR coupled to a 102 mm, f/8 refractor

8
Telescopes and Their Accessories for Use in Astroimaging

Any telescope can, of course, be used with a camera to photograph the heavens, but there is a problem that affects all telescopes that are not specifically designed for imaging. This is called 'curvature of field'. The point of focus of a simple telescope is at a fixed distance from the central point of the lens or mirror. Thus, as one moves off-axis, the point of best focus gets slightly closer to the objective as the 'focal plane' is actually the surface of a sphere. As the focal length gets less, the curvature of this surface gets greater and so the problem increases when short focal length telescopes are used.

The effect when using the flat sensor of a camera is such that, if the central part of the field is precisely in focus, the outer parts of the image will be somewhat out of focus and may also suffer from other aberrations, such as coma. There are two approaches to tackling this problem. The first option is to buy a telescope, termed an astrograph, which is specifically designed to provide a flat, aberration free, field, while the second option is to buy appropriate corrective optics that are located in front of the imaging camera.

Astrographs

Astrographs can be based on most telescope types. All incorporate additional optical elements within the tube assembly to achieve a flat field and pinpoint stars over what is termed their 'image circle'. All have image circles that will cover an APSC sized sensor; many will cover a full frame sensor, and a few are able to cover the larger digital sensors in medium format cameras.

Refractors

These have the highest cost per inch of aperture of any telescopes, but can produce exquisite images both when imaging and also when used for visual observations. TeleVue produce the NP 101is and NP 127is refractors which employ an additional

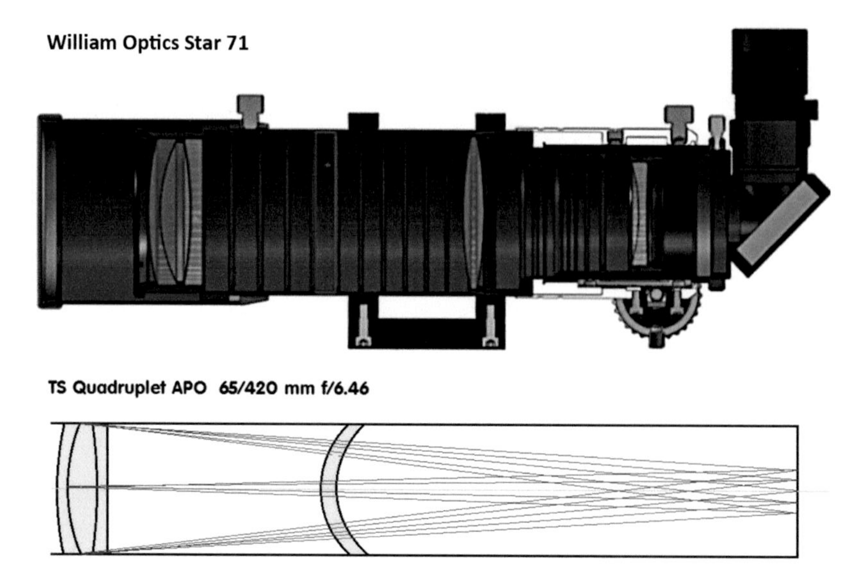

Figure 8.1 The optical configurations of the William Optics Star 71 and Teleskop Service 65Q astrographs

two-element lens within the tube assembly in what is termed a Petzval configuration. Takahashi produce the, f/5, FSQ-106ED Flatfield Super Quadruplet, which has a 530 mm focal length. Using a modified Petzval optical configuration this has an image circle of 88 mm – enough to cover up to 6 × 7 cm film or a 6 × 6 cm CCD sensor. They provide a focal reducer that reduces the focal length to 385 mm and the focal ratio to f/3.6. This still has an image circle of 44 mm – enough for a full frame 35 mm DSLR. Using the same optical design, they also produce the FSQ-85ED, which has an f/5.3, 450 mm focal length objective that can cover a full frame DSLR sensor. An optional focal reducer gives a focal length of 328 mm, a focal ratio of f/3.9 and a 40 mm image circle. At somewhat lower cost, William Optics recently introduced a patented five element design in their Star 71, f/4.9 astrograph, and Teleskop Service offer the 65Q astrograph which uses an air spaced triplet objective using one element of FPL-53 glass to provide excellent colour correction, along with a curved meniscus field flattener lens part way along the tube assembly. With a focal length of 350 and 422 mm, respectively, they both have image circles of 44 mm to allow the use of full frame DSLRs or CCD cameras.

Catadioptric Telescopes

Celestron have introduced their Edge HD series which, with apertures of 8 to 14 inches, incorporate lens systems within their baffle tubes to provide a flat field. The Edge HD's optical design provides tight, round, star images over a wide, 42 mm

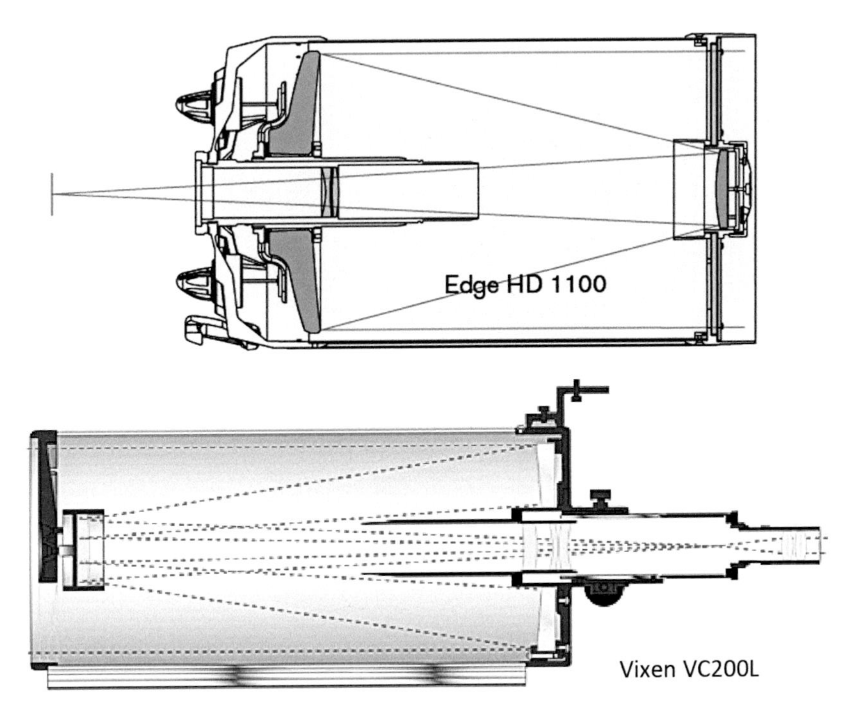

Figure 8.2 The optical configurations of the Celestron Edge HD and Vixen VC200L telescopes

diameter flat field of view so allowing full frame sensors to be used. Dedicated focal reducers are available for each telescope in the series. Mirror support knobs are used to lock the primary mirror in place and so prevent image shift during imaging. Similar corrective elements are used in the Vixen VC200L. This is a 200 mm aperture, f/9, catadioptric design which features a high precision sixth order aspherical primary mirror, a convex secondary mirror and a triplet corrector lens. The primary mirror is fixed (so there can be no mirror flop – an advantage when autoguiding) and focusing is carried out using a rack and pinion focuser at the rear of the primary mirror. A dedicated focal reducer is available to reduce the focal ratio to f/6.4 to give a wider field of view and allow shorter exposure times when imaging nebulae.

Corrected Newtonians

ASA in Austria make a range of corrected Newtonians with focal ratios of f/3.6, and Orion Optics in the UK provide Newtonians in their AG range which incorporate a Wynne corrector to provide a flat, coma free, field with focal ratios of f/3.8. These low focal ratios greatly reduce the required exposure times when imaging galaxies and nebulae and so make them perfect for deep sky astroimaging. They are also equipped with substantial focusers to support the weight of the imaging cameras.

Figure 8.3 The Orion Optics (UK) and ASA wide field corrected Newtonians

Corrective Optics: Field Flatteners, Focal Reducers/Flatteners and Coma Correctors

These are placed in the optical path of a telescope in front of the camera. A field flattener will do this with little, if any, effect on the focal length of the telescope. Two good examples are the Teleskop Service 2 inch Universal Field Flattener and the Hotech SCA Field Flattener, both of which will work with focal ratios from f/5 up to f/8.

As their name implies, focal reducers/flatteners will flatten the field, but in addition they reduce the effective focal length of the telescope – typically by ×0.7 or ×0.8. Many refractor manufacturers will provide specific versions to couple to their telescopes, while TeleVue provides a focal reducer/flattener that will work well with a refractor having a focal length of between 400 and 600 mm. Meade and Celestron provide focal reducers to convert their f/10 Schmidt–Cassegrains down to f/6.3.

Though not perhaps giving quite the performance of a Newtonian astrograph, a good compromise system is to couple a short focal ratio Newtonian with a coma corrector. Baader Planetarium produce the MPCC (Multi Purpose Coma Corrector)

Figure 8.4 The Andromeda Galaxy imaged by Peter Shah using an Orion Optics (UK) AG8 astrograph

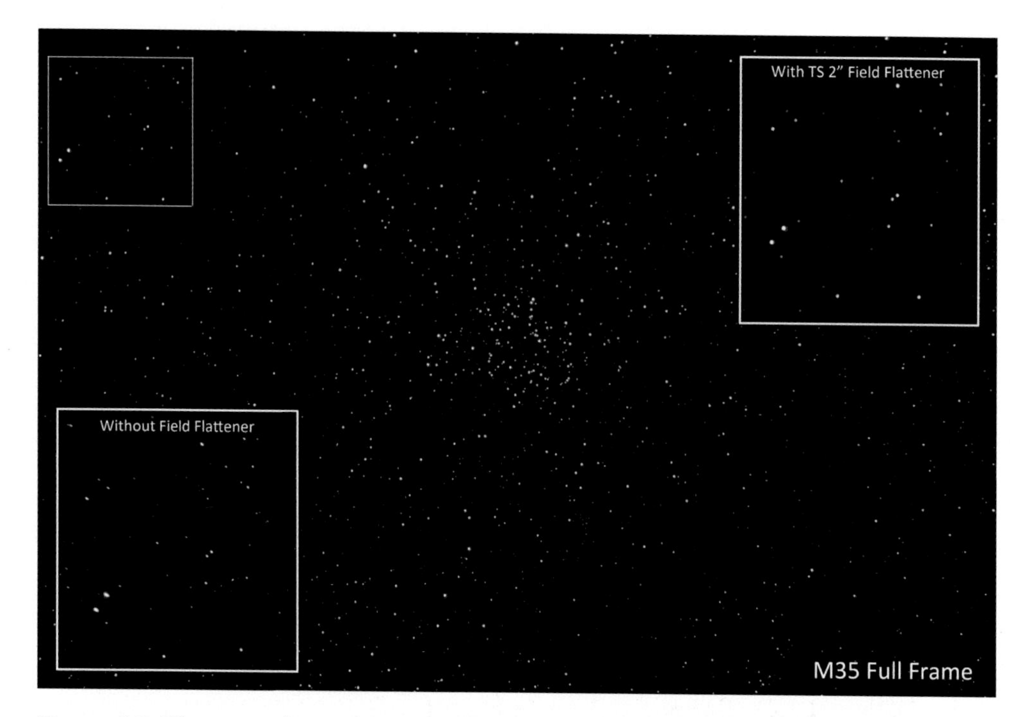

Figure 8.5 The open cluster M35 taken with an 80 mm ED refractor and Nikon D7000 DSLR showing the result of using a Teleskop Service 2 inch Universal Field Flattener to improve the image quality in the corners

Figure 8.6 TeleVue ×0.8 Focal Reducer/Flattener (left) and Baader Multi Purpose Coma Corrector (MPCC) (right) attached to DSLRs using their T-mount adapters

Mark III to cover a range of focal ratios and Altair Astro provide one specifically for 8 inch, f/4 Newtonians. SkyWatcher provide an f/4 Aplanatic coma corrector for their Quattro f/4 Imaging Newtonian. The image of the Leo Triplet shown in Chapter 20 was taken using an f/4 Schmidt–Newtonian using a Baader coma corrector.

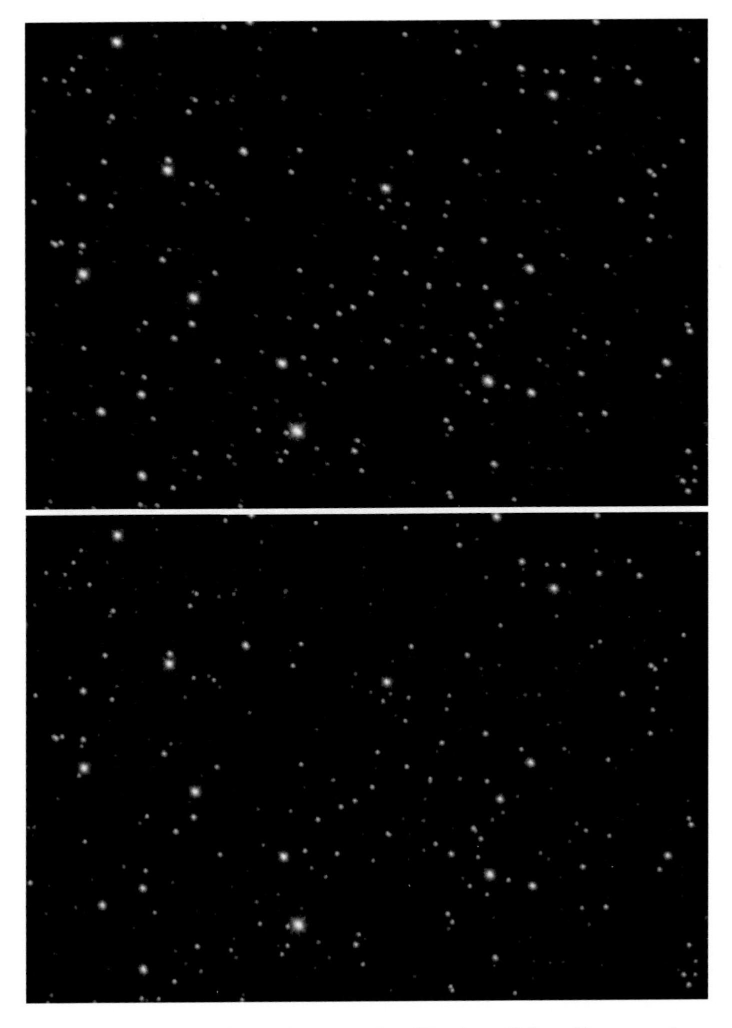

Figure 8.7 The effect of using the 'Darken' blending mode to correct distorted stellar images towards the corners of the frame

Taking 'Flat Fields' to Combat Vignetting

The use of focal reducers will often increase the amount of vignetting (darkening in the corners of an image). One good way of combating this is to use the telescope, *with the camera in the exactly the same orientation,* to take a number of what are called 'flat field' images of a uniformly illuminated target. The light in the sky at dusk is one possibility, while another is to place a diffusing sheet, perhaps a cotton sheet or a piece of translucent plastic, over the telescope aperture and aim towards a light source. The histogram should be inspected and the peak white should not exceed ~80 per cent of the full range of the sensor. These flat frames are input into *Deep Sky Stacker* along with the light and dark frames (when used) so that an appropriate correction can be made.

If an aftermarket accessory can provide good correction, why spend a considerable sum on an astrograph? It is a question of sensor size. If, as a beginner, one is using an APSC sized sensor, then the use of field flatteners, coma correctors or focal reducers will be fine, but if one advances to using full frame sensors then problems such as significant vignetting in the extremes of the field will begin to become apparent.

Correction Using Software

If, without the use of a field flattener or coma corrector, the stars in the corners of the frame are misshapen, then it may well be possible to improve them using software. In the frame corners the stars are usually elongated radially away from the centre. If this is the case, each corner *in turn* of the image is selected and the image duplicated to form a second layer. Then, in Layers, the 'Darken' blending mode is selected. The 'Move' tool symbol (top left in *Photoshop*) is then clicked on, followed by clicking in the image. The arrow keys are then used to move (pixel by pixel) the duplicated image over the background layer when, hopefully, more rounded stars will become apparent before the two layers are flattened. In the extreme corners more adjustment may be required and, if so, the process can be repeated with just that part selected. If the correction made by moving one pixel is found to be too much, the image size can be expanded by 200 per cent, the corrections made, and the image size reduced by 50 per cent back to the original size.

The author's image of the galaxy M33 in Triangulum, shown in Figure 21.8, was taken using a remotely operated ASA, 8 inch, f3.6 Newtonian astrograph located in Spain. The stellar images in the extreme corners of the frame attest to its superb, aberration free, flat field!

9
Towards Stellar Excellence

In a perfect astrophotograph, the stars should be perfectly round and the size of faint stars limited solely by the resolutions of the telescope and camera and the atmospheric seeing on the night in question. To achieve this, the tracking of the telescope mount should not waver by more than about 1–2 arc seconds as this is typically the limit set by the atmosphere. The tracking of all telescope mounts suffers from what is called 'periodic error' caused usually by imperfections in the worm drive of the right ascension axis. This can range from +/− 3 arc seconds for a top quality mount up to +/− 10 arc seconds or more for one of medium quality. This error repeats for each rotation of the worm and some mounts allow a periodic error correction (PEC) to be made by observing a star at high magnification and making manual tracking corrections over one cycle of the worm. These commands are stored in the drive computer and then applied automatically to reduce the tracking errors.

Another cause of imperfect stellar images is when the polar axis is not perfectly aligned on the North Celestial Pole. This causes the stars to 'trail', appearing like short sausages. Given *Adobe Photoshop*, the image can be duplicated and the 'Darken' blending mode selected in 'Layers'. By then moving the duplicated image pixel by pixel over the original image, the stars can be restored to their rightful shape. This will work when the stars are well separated but not when they are tightly packed as in the core of a globular cluster. In this case there are two solutions: the first is to reduce the exposure time down to perhaps 20–30 seconds, and the second is to use an autoguiding system which will allow longer exposures to be made that both reduces the amount of data to be handled and helps bring out fainter details. As my SBIG CCD camera takes 12 seconds to upload and store each image, longer exposures will also increase the effective imaging time. (If my exposures were just 12 seconds long, the effective imaging time would be halved!)

The autoguiding method that I, along with many others, employ is probably the easiest to use though more expensive than some other methods. A separate guide scope is mounted in parallel with the imaging scope on a very rigid crossbar (Figure 9.1). For the guide scope, I now use an 80 mm f/5 achromat sold by Rother Valley Optics which has very solid tube rings and an excellent focuser. (The rigid

Figure 9.1 An autoguiding set-up using a secondary guide scope and camera

crossbar and solid tube rings are important to eliminate any relative movement between the guide and imaging telescopes.) The short focal ratio gives a wide field of view, making it easier to find a suitable guide star. It is sometimes said that such a guide scope equipped with a sensitive guide camera will always find a guide star. This may well be true under dark skies, but I am not so sure that this the case under light polluted skies. It may then be necessary to offset the guide scope somewhat to find a suitable star. An Avalon Instruments X-Guider or SkyWatcher Guidescope Mount can be located between the crossbar and guide scope to allow it to be offset in either direction if needed (Figure 9.2).

Many guide cameras are available and for maximum sensitivity the choice should be for a monochrome one. Two of the best currently available are the Starlight Xpress Lodestar X2 and the QHY QHY5L-II (Figure 9.3). Both have peak quantum efficiencies (that is, the ratio between the number of electrons that are produced within a pixel well and

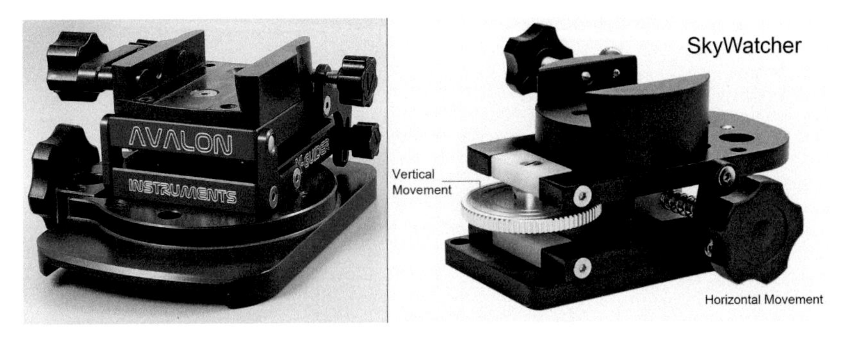

Figure 9.2 The Avalon Instruments X-Guider and SkyWatcher Guidescope Mounts

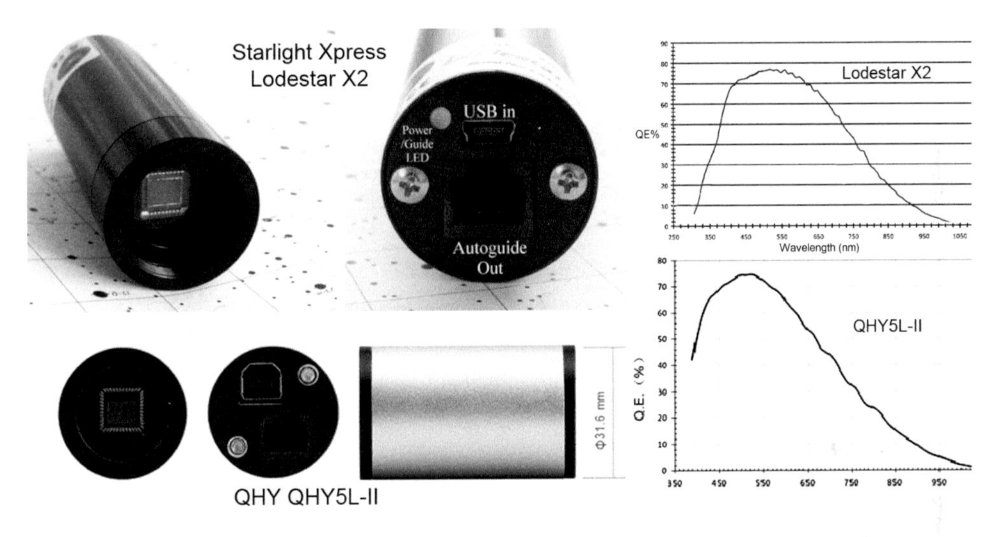

Figure 9.3 The Starlight Xpress Lodestar X2 and the QHY QHY5L-II guide cameras, showing their quantum efficiency plots

the number of photons falling upon it) greater than 75 per cent. The sensitivity falls off at both ends of the spectrum as is also seen in Figure 9.3. The guiding commands have to be passed to the telescope mount and, for this to be achieved, many mounts are equipped with an ST4 guiding port. Dedicated guide cameras such as those mentioned above will thus be equipped with an ST4 port whose output will transfer guide commands over a short cable to the mount. (These are received from the guiding software package using the USB connection from the computer to guide camera.) If the guide camera does not have a ST4 port, a Shoestring Astronomy USB Guide Port Interface can be used. Some mounts require serial guide commands from the computer, in which case a USB to serial converter will usually be needed.

The free guiding software used by many astrophotographers was written by Craig Stark and is called *PHD Guiding*, where PHD stands for 'push here dummy'. It is very

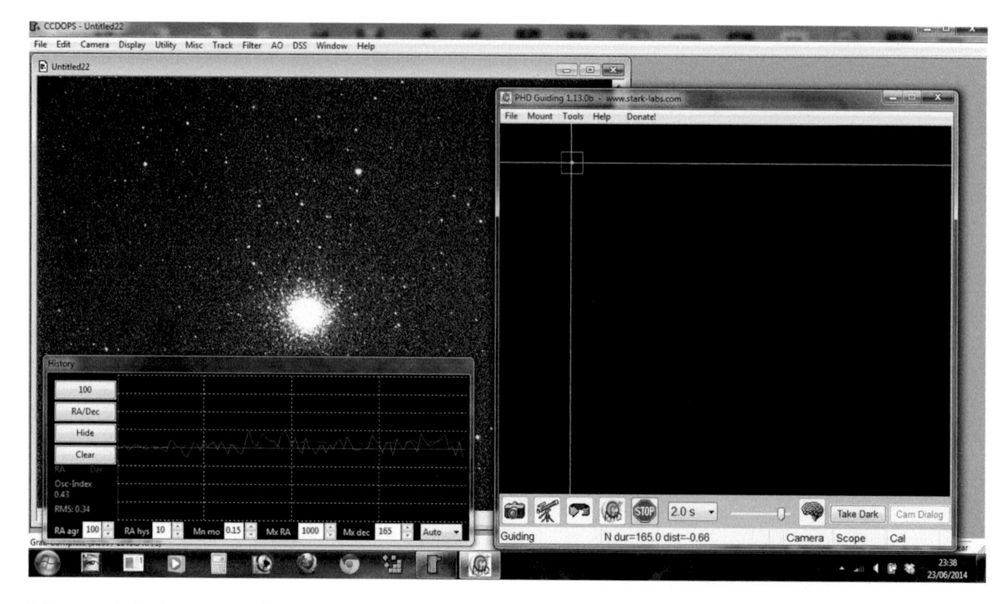

Figure 9.4 A screenshot showing, in the background, the stretched SBIG image of a single exposure of M13 and in the foreground the *PHD Guiding* window

simple to use, hence the name. The imaging telescope is centred on the target and a guide camera exposure time of typically 2–5 seconds selected. If necessary, the guide scope may be offset slightly to find a suitable guide star in the guiding camera's star field. The star is selected and the autoguiding routine entered. The program first sends out drive commands to the telescope in both axes and observes how the mount responds and, when calibrated after a few minutes, will then send suitable drive commands to keep the guide star stationary in the field, and imaging can begin.

As a first autoguiding test, I imaged the globular cluster M13 in Hercules using a 127 mm refractor and SBIG ST-8300 imaging camera mounted in parallel with the guide scope and camera on a Losmandy GM8 equatorial mount, as seen in Figure 9.1. (As this chapter is about guiding rather than about cameras, I have not expanded on the use of the SBIG cooled monochrome CCD camera, which is covered in depth in Chapter 20.)

In the background of the screenshot shown in Figure 9.4 is a single 60-second exposure of M13 produced by the SBIG camera. (It is *not* overexposed; the capture software greatly stretches the displayed image but not, of course, the raw data, which is saved as a 16-bit TIFF file.) In the foreground is the *PHD Guiding* window, which shows the star field imaged by a QHY QHY6 guide camera with the selected star pinpointed, and a plot showing how well the guide star remains stationary in the field, indicating how well the mount is tracking. The plot showed that the rms (root mean square) error was 0.34 pixels with a peak error less than one pixel of the guide camera sensor. This translated to a peak tracking error of 1.5 arc seconds in the image produced by the SBIG camera.

Figure 9.5 The *Photoshop* technique to remove light pollution from an image

Removing Light Pollution from an Image

Thirteen 1-minute exposures were taken and stacked in *Deep Sky Stacker*. The image was processed and the light pollution background removed using *Adobe Photoshop*. First, some stretching was applied to bring out the fainter stars (Figure 9.5, left), then the image was duplicated and the 'Dust and Scratches' filter applied to the duplicate layer with a pixel radius of 12 pixels (Figure 9.5, middle). This left the globular cluster and two bright stars still visible, which were then cloned out from neighbouring parts of the image to give an image of the light pollution (Figure 9.5, right). The two layers were then flattened using the 'Difference' blending mode to remove the light pollution.

A little local contrast enhancement was made using the 'Unsharp Mask' filter with a large radius and small amount (adjusted by trial and error) to give the final image of M13. The very high overall contrast of the telescope allied to the excellent tracking of the mount has enabled the structure of the inner core of the cluster to be visible. Without autoguiding, this would quite likely have been blurred out. Other methods of autoguiding and more details of the techniques involved are given in Appendix D.

Longer total exposure times failed to bring out further fainter stars in the outer parts of the cluster and this was due to the masking effects of light pollution. The use of an Astronomik CLS light-pollution filter (discussed in Chapter 12) would certainly help, as would, of course, imaging the cluster from a dark sky location.

Figure 9.6 The globular cluster M13 imaged from a light polluted location using a 127 mm refractor and SBIG ST-8300 monochrome CCD camera

10
Cooling a DSLR Camera to Reduce Sensor Noise

Digital cameras are capable of superb results in winter when the ambient temperature is low, but tend not to produce such good images during the summer. This is simply due to the fact that the warmer the camera sensor, the more thermal electrons are generated (dark current), so affecting the images. The program *Astro Photography Tool*, for use with Canon cameras, has the useful facility of appending the sensor temperature to the filename of each raw image. Observing how the temperature increases as a continuous sequence of long exposure images is taken is quite instructive. With my Canon EOS 1100D camera it rises over a period of ~45 minutes and then stabilises at a temperature of about 12° Celsius above the ambient temperature, as shown in the two examples of Figure 10.1.

The dark current approximately doubles for each 6 degree Celsius rise in sensor temperature, so this will increase by about four times as the sensor warms up. On warm nights in summer the sensor temperature could easily exceed 25° Celsius. One effect of the dark current is to fill the pixel wells with thermal electrons rather than those due to the incident light. The result is that with long exposures the available 'depth' for the image is reduced and so the dynamic range of the sensor is reduced. (This is one area where film is better – there is nothing comparable to dark current.) The second effect is to make the images somewhat noisier, and this increases as the square root of the total dark current, so a 12°C rise in sensor temperature will double the dark current noise in the image.

The temperature at which the sensor will stabilise will depend on the thermal path to the rear of the camera. This will be less if the camera has a vari-angle rear screen which can be moved away from the camera back. One might even consider mounting a small heatsink on the back panel using thermal paste!

A very low cost system that can provide a useful amount of cooling in warm weather is to simply mount the camera in a plastic box sufficiently large (I used a 2.3 litre food box) so that the camera can be surrounded by small flexible ice packs and insulation. Food containers have lids designed to provide an air tight seal, which is

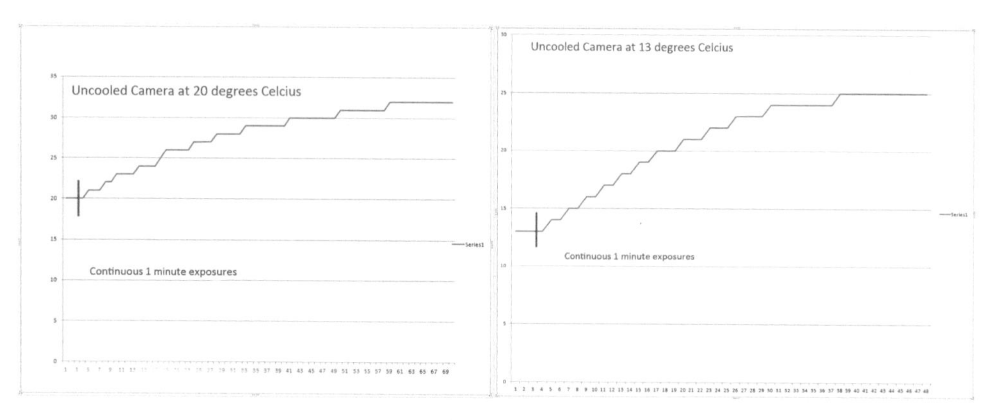

Figure 10.1 The increase in sensor temperature as continuous exposures are taken, both examples giving a rise of 12° Celsius above the ambient temperature

Figure 10.2 The icepack coolbox: (left) the foil and gauze insulation; (middle) the camera installed with three icepacks; (right) the final icepack to place behind the camera with insulated lid

just what is wanted. An excellent insulation material can be made by layering pieces of aluminium foil with a fine mesh such as net curtain material, which is placed around the insides of the box. The silver foil reduces heat transfer by radiation and the many air pockets produced between the layers reduce heat transfer by convection (Figure 10.2, left). Loosely packed layers of kitchen paper (which include air pockets) and segments of expanded polythene foam can be used to fill any remaining space. (Note, the plastic used for the main part of these boxes is very brittle, so the hole at the front for the T-mount to pass through is best made by using a hot skewer to burn round the edge of the hole.)

It is, of course, necessary to remotely control the camera through a USB cable. For Canon cameras the programs *Astro Photography Tool* or *BackyardEOS* will achieve this, while for Nikon cameras a new program, *BackyardNIKON*, has become available. The program, *Digicamcontrol*, can also remotely control many Canon and Nikon DSLRs. As batteries produce some heat and may not last too long in cold conditions, I use an externally powered Canon AC adaptor kit.

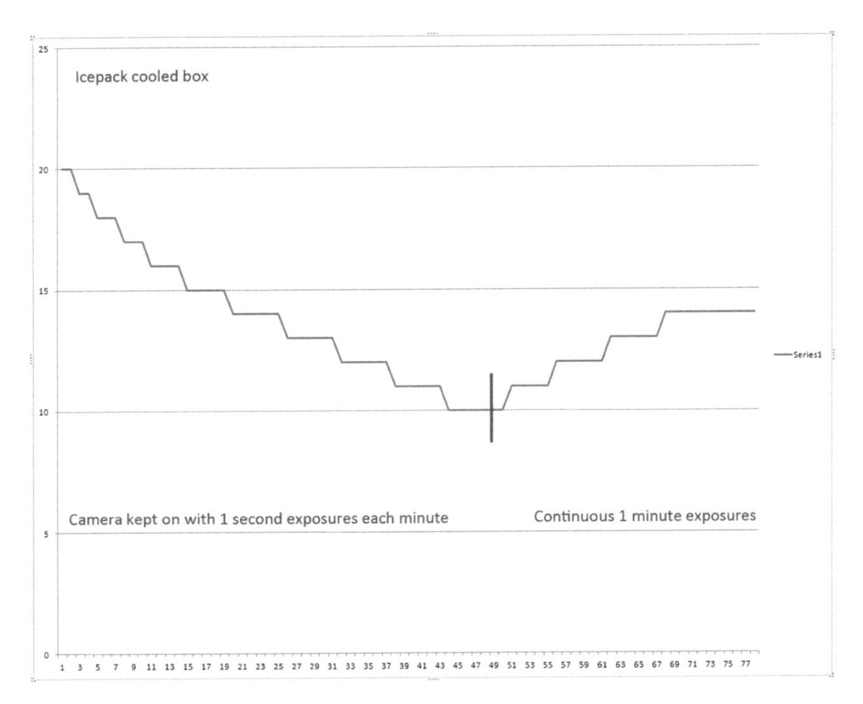

Figure 10.3 The sensor temperature when the camera is cooled at an ambient temperature of 20 degrees Celsius using the icepack coolbox

To pre-cool the camera, it is placed within a plastic bag (along with a desiccant pack to prevent moisture condensing on the camera), and put in a refrigerator for around 20 minutes. It is then mounted (in a thin plastic bag along with some desiccant packs with a suitable hole for the T-mount to pass through) in the cool box with three ice packs placed around it (Figure 10.2, middle). The camera is switched on, a final ice pack is placed directly behind the camera back and the insulated back cover is attached (Figure 10.2, right).

There is one problem that must be addressed when cooling a DSLR camera. If the air is humid, then moisture could easily condense on the sensor. To prevent this, the amount of air in contact with the sensor must be minimised. This can be achieved by screwing in a 2 inch filter – perhaps a UV cut-off filter – at the front of the T-mount. The use of a field flattener or focal reducer will have the same effect.

Using *Astro Photography Tool* to control a Canon camera, short, one-second exposures are then taken every minute (so keeping the camera on) whilst monitoring the sensor temperature (given in the real time log) as the sensor cools. In one test made at an ambient temperature of 20° Celsius, I allowed the sensor temperature to drop to 10° Celsius as seen in Figure 10.3 and then took a continuous sequence of 1 minute exposures. The sensor temperature increased rapidly to 14° Celsius where it stabilised and remained within one degree for well over two hours (as opposed to ~32° Celsius without cooling).

Figure 10.4 A comparison of the sensor noise at 14° Celsius (left) and 32° Celsius (right)

The 18 degree reduction produced by the simple cooling jacket would reduce the dark current by a factor of ~8 and the sensor noise by the square root of 8 or 2.8 times – well worth having. As can be seen in the identically stretched images of Figure 10.4, the dark current noise (which will be present in any image) is significantly greater at 32° than at 14°C. I find it interesting that it is the red component that appears to have increased the most.

The effect of using a cooling box and ice packs will give a stabilised temperature (sensor temperature ~14° Celsius) irrespective of the ambient temperature. If this is lower, the temperature will be stabilised for longer as the heat gained from the surrounding air (so melting the ice packs) will be reduced. Of course, if the ambient temperature is close to zero degrees Celsius or less, there would be no point in using a simple cooling box.

Employing a Peltier Cooling System to Reduce Thermal Noise

It is possible to buy a Geoptik thermoelectric cooling box from Teleskop Service in Germany for use with the smaller Canon EOS cameras such as the EOS 350, 400, 500, 1000, 1100 and similar sized models but not for full frame models such as the EOS 5D. As shown in Figure 10.5, this weighs 1 kg, so could be supported with the included camera by most quality focusers. It claims to reduce the internal temperature of the unit by up to 20° Celsius in 30 minutes. A display shows both the internal and ambient temperatures. It requires a 5 amp, 12 volt supply provided through a cable and cigarette lighter plug.

The Italian company Primaluce Lab sells a modified Canon EOS 700D having an 18 megapixel, APSC sized sensor modified to be more sensitive to the light of H-alpha emission. Shown in Figure 10.6, this uses a two stage Peltier set point cooling system which can cool the sensor down to a defined temperature up to 30° Celsius below the ambient temperature. This allows dark frames to be taken at the same temperature for calibration purposes. Importantly, it includes an internal anti-dewing system.

Figure 10.5 The Geoptik thermoelectric cooling box

Figure 10.6 The Primaluce Lab, Peltier cooled, Canon EOS 700Da

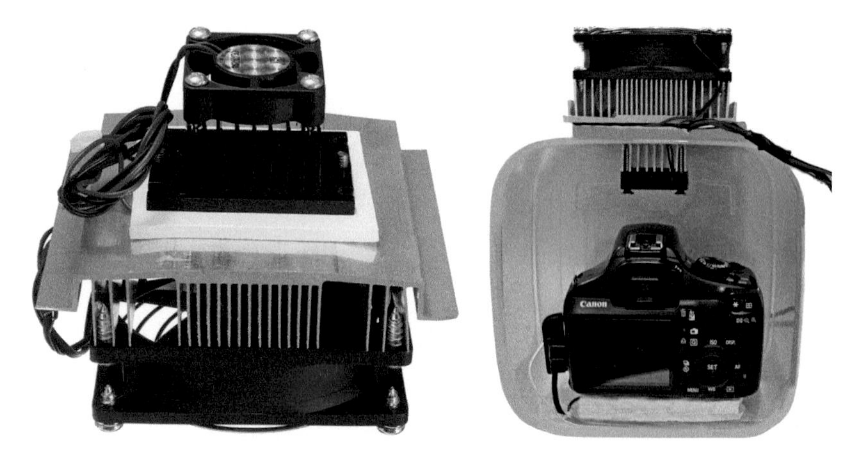

Figure 10.7 The Peltier cooling kit (left) and (right) installed (without insulation) in a 2.3 litre food box

Home Constructed Peltier Cooling Systems

If one searches the Internet, one can find several references to home built Peltier cooling systems, such as that designed by Gary Honis and described at http://dslrmodifications.com/rebelmod450d16c.html. One can obtain Peltier cooling kits as shown in Figure 10.7 (left) from e-Bay and I have been able to purchase one at very low cost which employs a 72 watt Peltier unit requiring 6 amps at 12 volts to power it – Peltier coolers are *very* inefficient! To make a very simple Peltier cooled box, a similar 2.3 litre coolbox to that used in the icepack cooling system was employed with a second opening placed at the top so that the small heatsink and fan on the cold side of the Peltier unit can extend into the coolbox. The camera location must thus allow for its presence and be placed nearer to the bottom of the box as shown in Figure 10.7 (right).

A cooling test at an ambient temperature of 14° Celsius, shown in Figure 10.8, gave a stabilised sensor temperature of 8° Celsius and thus, again, reduced the sensor temperature by 18 degrees. (It would have been ~26° Celsius without cooling.) Given a suitable power supply, the advantage over the icepack version is that the sensor temperature can stay low for as long as desired.

Peltier coolers can be also usefully employed in the winter months as they then have the ability to reduce the temperature of the cooling box to well below zero and hence bring the sensor temperature close to zero Celsius. Comparing this to an uncooled camera in summer when the sensor temperature might well be 30° Celsius, this would reduce the dark current by ~32 times and hence reduce the dark current noise by over 5 times.

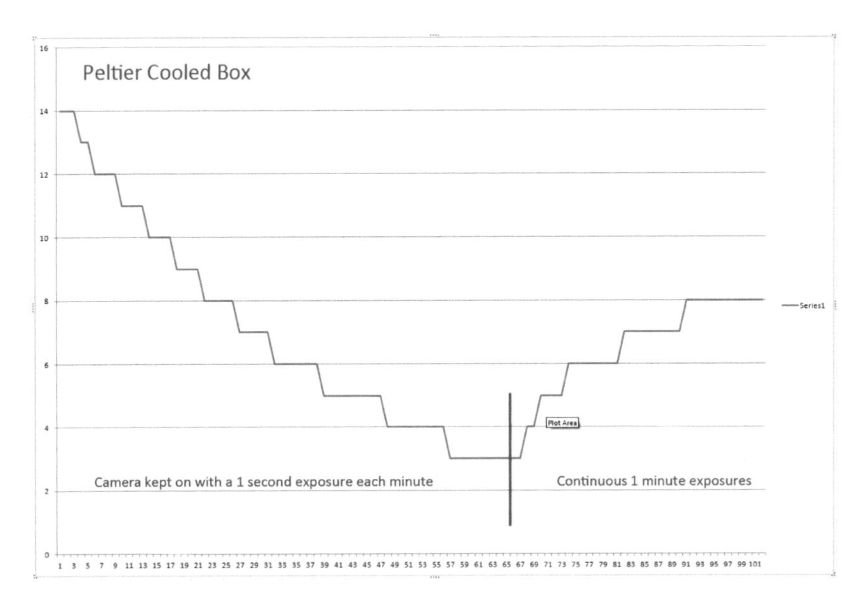

Figure 10.8 The sensor temperature when cooled by the Peltier unit at an ambient temperature of 14° Celsius

11 Imaging the North America and Pelican Nebulae

The region near Deneb in Cygnus containing the North America and Pelican Nebulae is a perfect late summer target for a DSLR coupled to a short focal length refractor. The North America Nebula (NGC 7000) is a region of nebulosity in which hydrogen gas, excited to glow in the deep red colour of the H-alpha line, forms the outline of North America, with a dust cloud darkening the region that forms the 'Gulf of Mexico'. To its upper right is a further region of nebulosity in the shape of a Pelican's head.

As described in Chapter 7, in order to capture the H-alpha emission well, a Canon 60Da, a Nikon 810A or a Canon camera modified to be more sensitive to H-alpha emission should ideally be used. For my attempt at imaging the nebulae, a modified Canon 1100D DSLR was attached to a William Optics 72 mm Megrez ED refractor using a T-mount and Teleskop Service, 2 inch field flattener. The camera (through a USB cable) was under the control of a program called *Astro Photography Tool* running on an adjacent laptop. This allowed a sequence of 138 raw frames to be taken at 800 ISO using an exposure of 30 seconds to eliminate any tracking errors from the unguided equatorial mount, so giving a total exposure time of 69 minutes.

These raw frames were imported into *Deep Sky Stacker* to be aligned and stacked. The result, as usual with *Deep Sky Stacker* when imaging nebula, was initially very disappointing as nothing of the nebula was visible in the resulting TIFF file. To bring out the faint nebulosity, the image must be 'stretched'. Without the use of *Adobe Photoshop*, the free program *IRIS* can be used: the TIFF file is loaded and in the 'View' menu, 'Logarithm' chosen. By adjusting the two sliders, the nebulosity can be made to appear and, if saved as a JPEG, this can then be enhanced in an image processing program. A better result may result from using the 'Curves' function in *Photoshop*. The image is opened and the curve shown in Figure 11.1 is applied several times – so making this an Action is useful. This curve does not affect the brighter stars but lifts up the faint nebulosity, giving the result shown in the image.

If the imaging, as in this case, has taken place in a light polluted location, the image will have an overall red cast. A way to remove this, given the facility to use layers in *Photoshop*, is to first duplicate the layer, then use the colour picker to select the colour visible in the darkest part of the image in the 'Gulf of Mexico'. As there

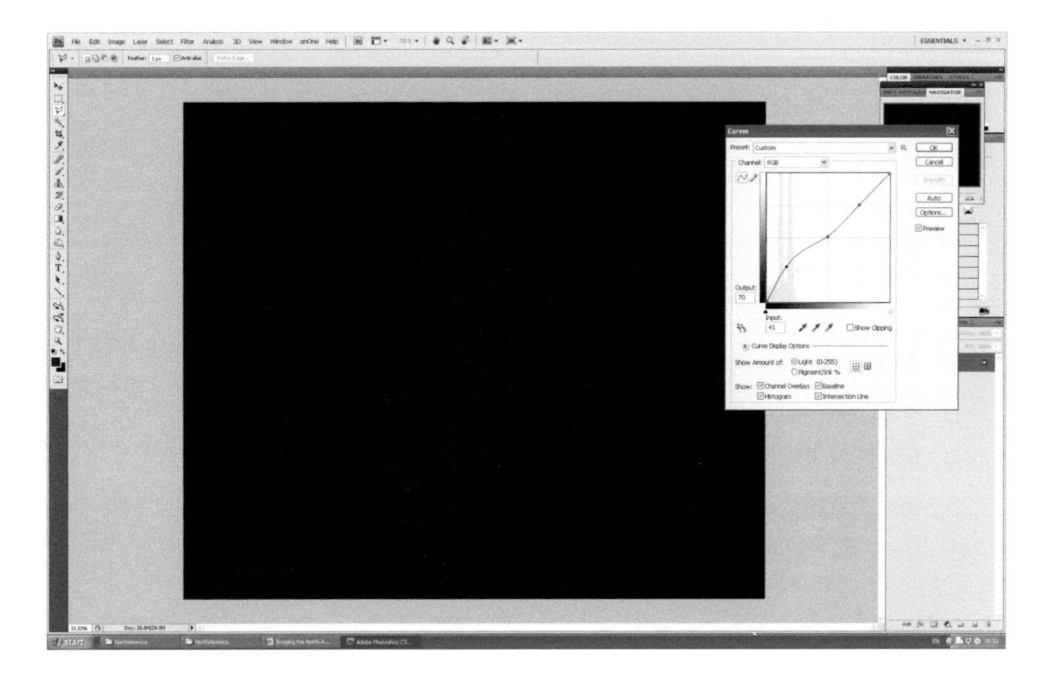

Figure 11.1 The result of stretching the output from *Deep Sky Stacker* using the 'Curves' function in *Adobe Photoshop*

is little H-alpha emission visible in this region owing to a thick dust cloud, this deep red colour is representative of the light pollution across the whole image. (It helped in this instance that Cygnus was high in the sky, so the light pollution could be expected to be uniform across the image.) The whole of the duplicate layer is then painted over with this colour as seen in Figure 11.2, the blending mode is set to 'Difference' and the two layers flattened. This removes the overall red cast and only the faint outline of the nebula remains visible.

Further applications of the same Curves function were applied with an adjustment of the black level in 'Levels' to give the result seen in Figure 11.3. This was then saved as a TIFF file called 'stars+nebula'.

There is a technique, first described in Chapter 3, to enhance the visible structure in the nebula but which tends to make the stars more prominent. So a useful technique is to remove the stars from the image, enhance the nebula and then put the stars back. The 'Dust and Scratches' filter is applied to the 'stars+nebula' image with a radius of ~12 pixels and, as the filter regards the stars as dust, they magically disappear. Some of the brighter stars that remain can be cloned out from adjacent parts of the image. This image should be saved as 'nebula' and is shown in Figure 11.4.

The original 'stars+nebula' image is brought back, copied and pasted over the nebula image and, with the difference mode selected, the two layers flattened. This just leaves the stars visible ((stars + nebula) – nebula = stars). These may well have a

Figure 11.2 The 'light pollution' layer derived from the dark region of the 'Gulf of Mexico'

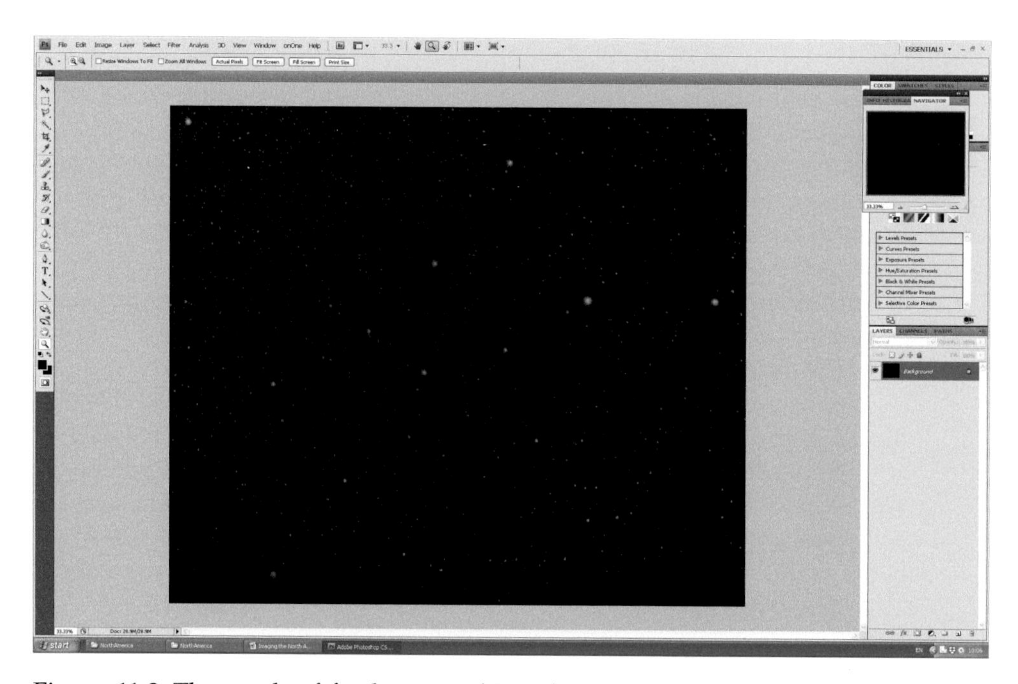

Figure 11.3 The result of further stretching the image, having removed the light pollution

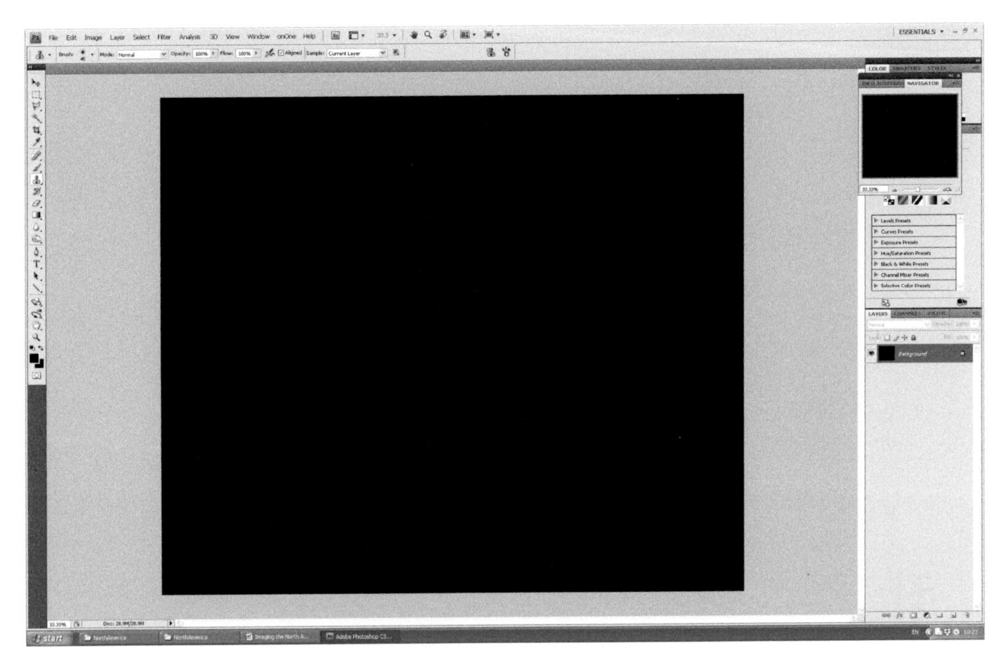

Figure 11.4 The 'nebula' image, having removed the stars from the image using the 'Dust and Scratches' filter in *Photoshop*

green cast, and if so, this can be removed with a free 'plugin' for *Photoshop* written by Rogelio Bernal Andreo and called 'Hasta La Vista Green': search for 'Deep Sky Colors'. Once installed in the 'Plug-ins' folder in the *Photoshop* file directory, this will appear as a filter within 'Deep Sky Colors', and simply applying the filter removes the green cast. The stars image should then be saved as 'stars' as shown in Figure 11.5. If the star images appear somewhat bloated, the *Photoshop* filter 'Minimum' (Filter > Other > Minimum) can be applied to the 'stars' image. The effect of the Minimum filter, even with only one pixel selected, is often too great and can be reduced by increasing the image size to 200 per cent, applying the filter and then reducing it by 50 per cent to return to the original size.

Local contrast enhancement can then be applied to the nebula image using the 'Unsharp Mask' filter in *Photoshop* (Filter > Sharpen > UnsharpMask) with the radius set initially to a large value, say 200, and the amount ~20–60 adjusted to bring out the nebula structure. This is a matter of trial and error. It can then be selectively applied to smaller regions of the nebula, for example the 'Cygnus Wall' region towards the bottom and the 'Pelican Nebula' to the upper right of the image. One simply experiments until a pleasing nebula image results, as shown in Figure 11.6.

The final step is to bring the stars back: the 'stars' image is opened, copied and pasted over the 'nebula' image and the 'Screen' blending mode selected to bring the stars back into the image. As the image is about the nebula, not the stars, the brightness of the stars can be reduced if desired by adjusting the opacity slider downwards somewhat. The

Figure 11.5 An image containing just the stars derived from the images of Figures 11.3 and 11.4

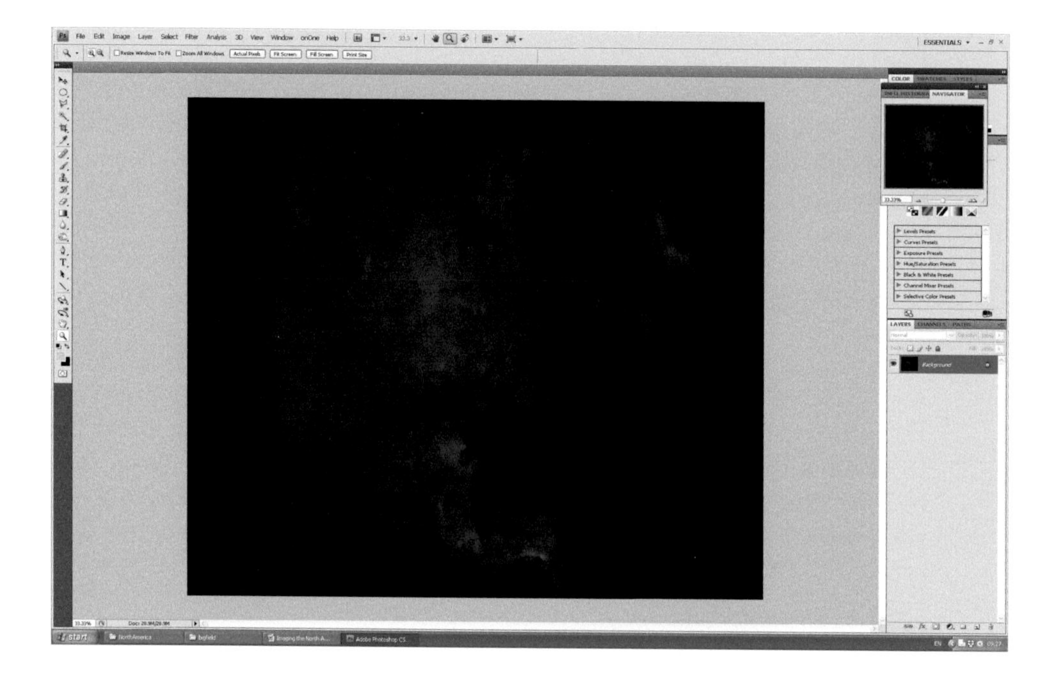

Figure 11.6 The result of enhancing the nebula using a local contrast function

Figure 11.7 The North America Nebula and Pelican Nebula imaged with a William Optics 72 mm Megrez refractor and H-alpha modified Canon 1100D DSLR

result is the final image of the North America Nebula and Pelican Nebula shown in Figure 11.7. The North America Nebula region is an object that requires more processing of the initial stacked image than that of any other celestial object I know of. The several techniques that have been used to reach the final image can be applied to any of your astronomical images and will, I hope, help you to achieve more pleasing images yourself.

12
Combating Light Pollution – the Bane of Astrophotographers

For many of us who live in towns and cities, light pollution can be a real problem. This chapter aims to describe some techniques that can be used to mitigate its effects, using filters and software. Light pollution is a particular problem when attempting to image faint nebulae. The problem is that if the surface brightness of the nebulosity, or even faint stars, is significantly less than that of the light pollution, then increasing the exposure time will not be able to bring them out. Figure 12.1 is a comparison of an image I made using a 5 inch refractor of the globular cluster M13 under the light polluted skies at my home, along with an image derived from the ESO (European Southern Observatory) Deep Sky Survey. Increasing the total exposure time did not enable me to bring out any of the fainter stars that are seen in the outlying parts of the cluster in the ESO image. These faint stars simply cannot be imaged under heavily light polluted skies.

A second problem caused by light pollution is that if long exposures are used, the sensor will saturate with its light and so overwhelm the image being captured. Thus the maximum allowable exposure times will be significantly reduced.

The obvious solution is to image from a dark sky site, but this may not be always possible. If not, what can be done? The first thing to note is that the level of light pollution will vary considerably depending on the amount of dust and aerosols in the atmosphere. On some nights, sadly few in number, the atmosphere is very clear and is said to have high transparency (see Appendix C). Little light is then scattered back towards the Earth. When the transparency is low, not only is more light scattered back to Earth, but the light from the stars and nebulae is attenuated as well – a 'double whammy'! So the first piece of advice is to image when the sky is transparent. Even from my home, not far from the centre of a small town, there are a few nights when I can see magnitude 4.5 stars overhead and just make out the arch of the Milky Way.

Depending on the type of object being imaged, light pollution filters may help. An ideal case for the use of filters is when much of the light from the object is emitted at a few specific wavelengths. One good example is the planetary nebula M27, the Dumbbell Nebula. The gases in the surrounding clouds are excited by ultraviolet light emitted by the white dwarf star at the heart of the nebula. Hydrogen gives

Figure 12.1 The effect of light pollution masking the fainter stars in M13, as described in the text
Images: Ian Morison and ESO Deep Sky Survey.

rise to the deep red H-alpha and blue-green H-beta lines, while oxygen gives rise to the green O III line. Placing a Baader UHC-S or, even better, the Hutech IDAS LPS-V4 (which has a narrower pass band at the H-alpha line) in front of the sensor will significantly reduce the light pollution but allow the key wavelengths emitted by the nebula to pass.

Imaging galaxies or star clusters, whose light covers much of the visible spectrum, requires a different kind of filter. This type aims to reduce the light from the main light pollution causes such as sodium or mercury lamps, but allow as much of the remaining light to pass as possible. Examples are the Astronomik CLS filter, the Orion SkyGlow Imaging Filter and the Hutech IDAS LPS-P2.

The upper part of Figure 12.2 shows the full solar spectrum and the passbands of the Astronomik CLS, Baader UHC-S and Hutech IDAS LPS-V4 filters along with the location of the main emission lines. Below are shown strips of the sky imaged from my home location under low cloud (so effectively imaging the light pollution) with the 'sky' strip having no light pollution filter in line and the strips below having the Astronomik CLS, Baader UHC-S and the Hutech IDAS LPS-V4 filters placed prior to the camera. All exposures were identical. The Astronomik filter seems particularly effective and its use might well allow me to image fainter stars in the M13 globular cluster. However, the character of light pollution is changing as 'orange' sodium street lamps are being replaced by LED lamps. Their design is good in that they send less light skywards, but as they emit 'white' light, the use of a CLS or UHC-S filter may well not be so effective in future.

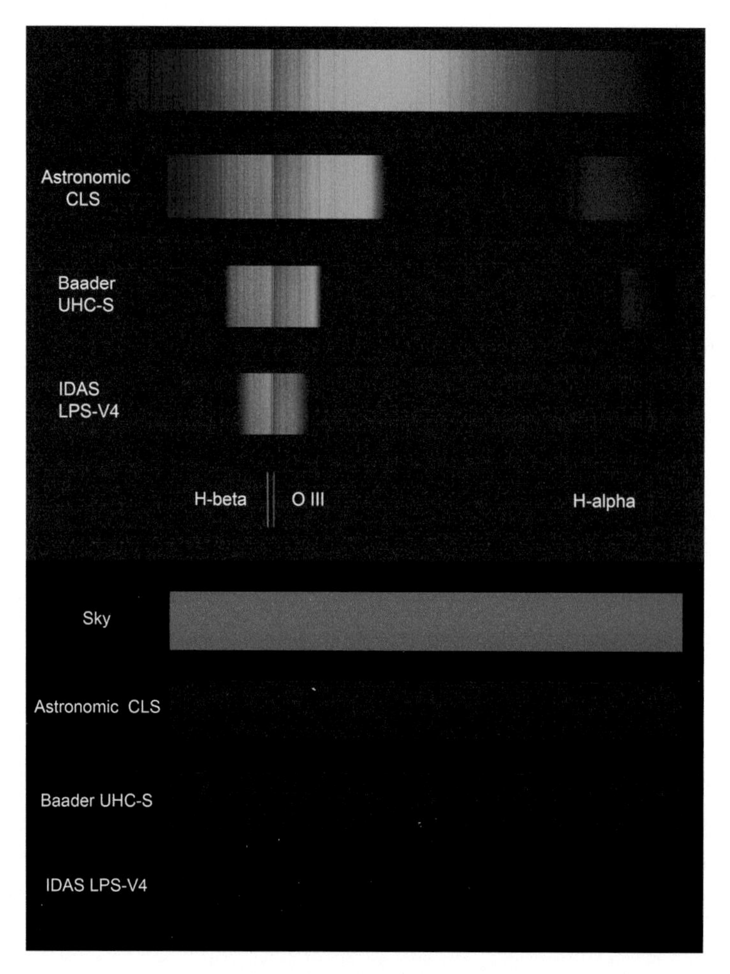

Figure 12.2 The passbands and effect of using three light pollution filters

As an example of the use of a light pollution filter, two images of the Dumbbell Nebula, M27, were taken using a Canon EOS 1100D (modified to increase its sensitivity to H-alpha emission) coupled to a 200 mm aperture, f/9, Vixen VC200L telescope, which provided a good sized image of the nebula on the camera sensor. Forty-seven 1-minute exposures at ISO 800 were made and, as described in previous chapters, and stacked in *Deep Sky Stacker*, whose output TIFF file was 'stretched' to bring out the nebula. This also made the overall red cast of the light pollution very apparent. The upper half of Figure 12.3 shows the result derived without use of a light pollution filter, while below is the image when a Baader UHC-S filter was incorporated. The effect is obvious.

In either case, the light pollution can be removed by using a technique in *Adobe Photoshop* that was applied to the image taken using the Baader filter. This is a three step process. As shown in Figure 12.4, the image layer is duplicated to

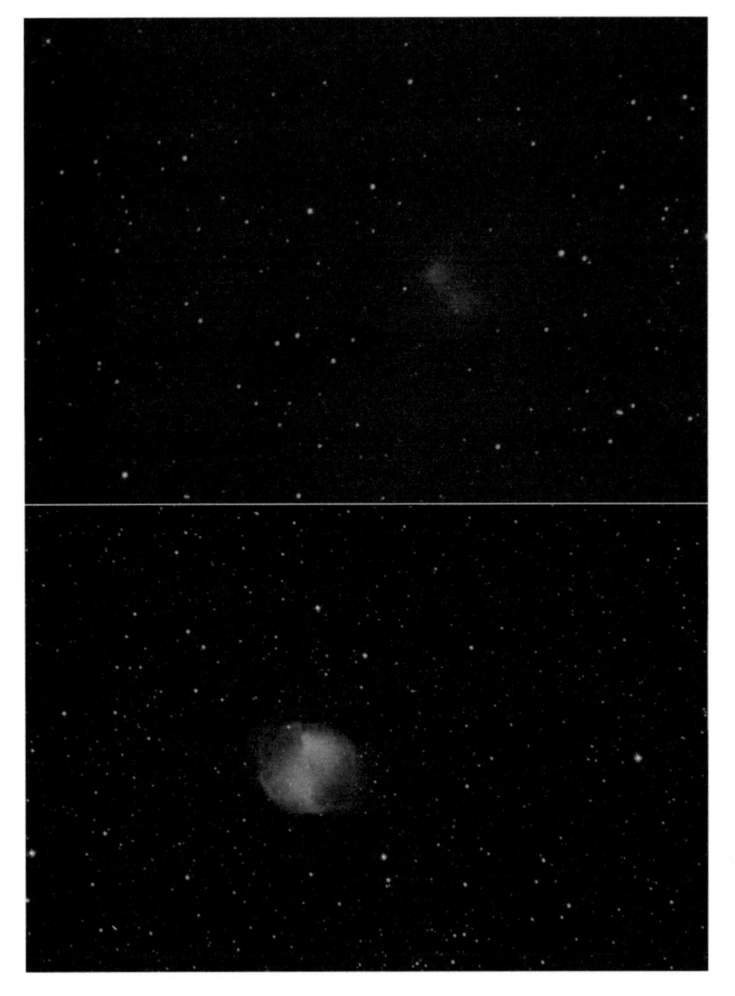

Figure 12.3 The reduction in light pollution in an image of M27, the Dumbbell Nebula, when using a Baader UHC-S light pollution filter as shown in the lower image

give a second copy and the 'Dust and Scratches' filter applied to the duplicate layer with a radius of ~12 pixels. As if by magic, the majority of stars disappear but the nebula still shows through.

In Figure 12.5, the nebula has been cloned out from the adjacent areas of the image and a Gaussian Blur applied to give a very smooth image of the light pollution.

Figure 12.6 shows the result of selecting the 'Difference' blending mode and flattening the two layers to remove the light pollution.

(Note: In attempting to remove the light pollution from a colleague's heavily light polluted image of a comet, I found it best to only remove part of the light pollution layer by reducing the opacity to 40 per cent before differencing the two layers. The process was then repeated twice, so gradually removing the light pollution.)

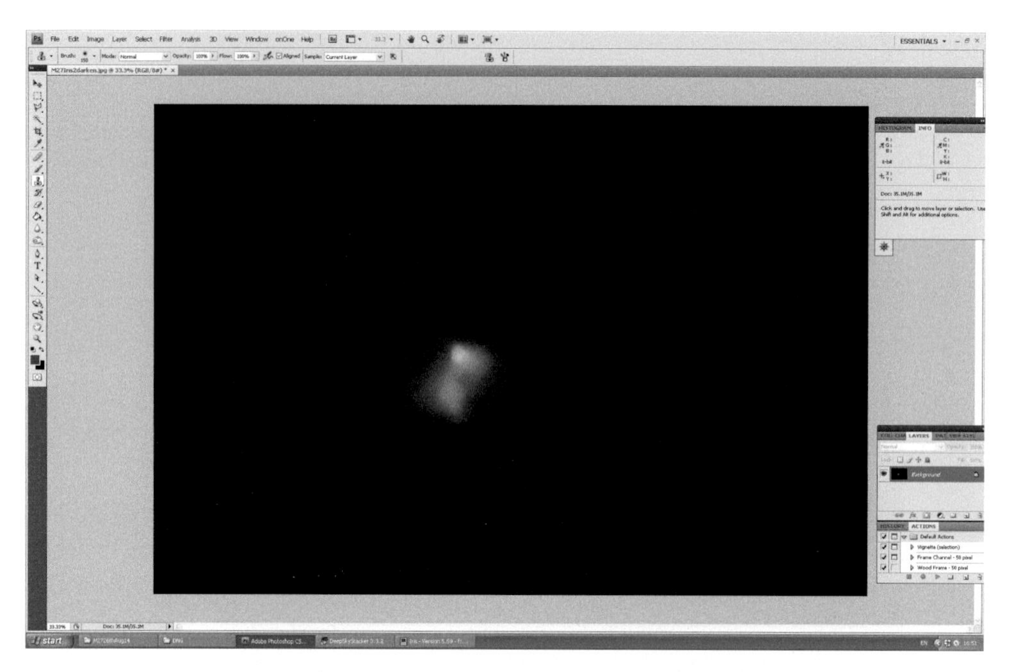

Figure 12.4 Having applied the 'Dust and Scratches' to a duplicate image of the filtered M27 image. The nebula shows through the light pollution

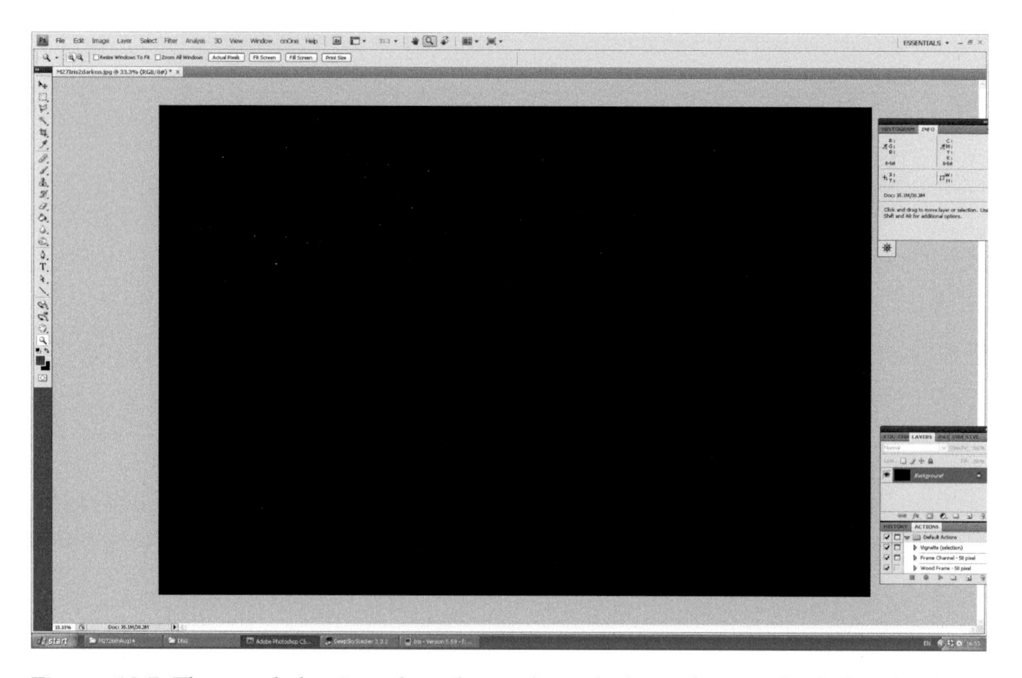

Figure 12.5 The result having cloned out the nebula and smoothed the duplicate image with the 'Gaussian Blur' filter to give an image of the light pollution

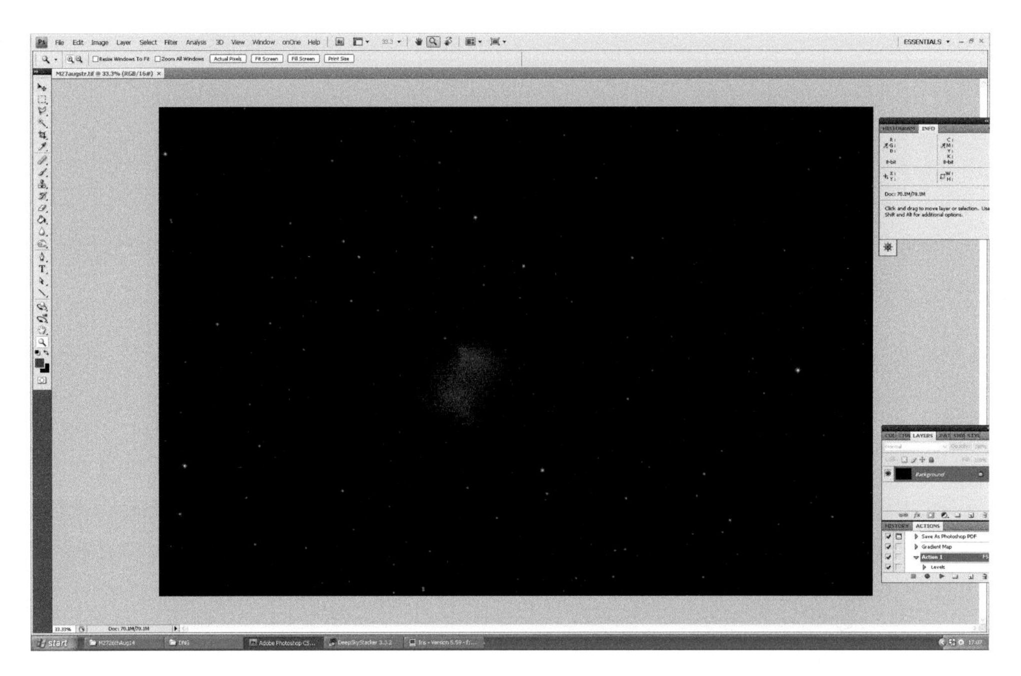

Figure 12.6 The result of differencing the two layers to remove the light pollution

Two further image enhancement techniques were applied to achieve the final image shown in Figure 12.7. First, the area including the nebula was selected and some local contrast adjustment applied using the 'Unsharp Mask' filter with a radius of ~200 and amount of ~20 per cent. (If this process were applied to the whole image, the stars would become too obvious.) With the same mask selected, a little sharpening was applied using the 'Smart Sharpen' filter to give the final cropped image of the Dumbbell Nebula.

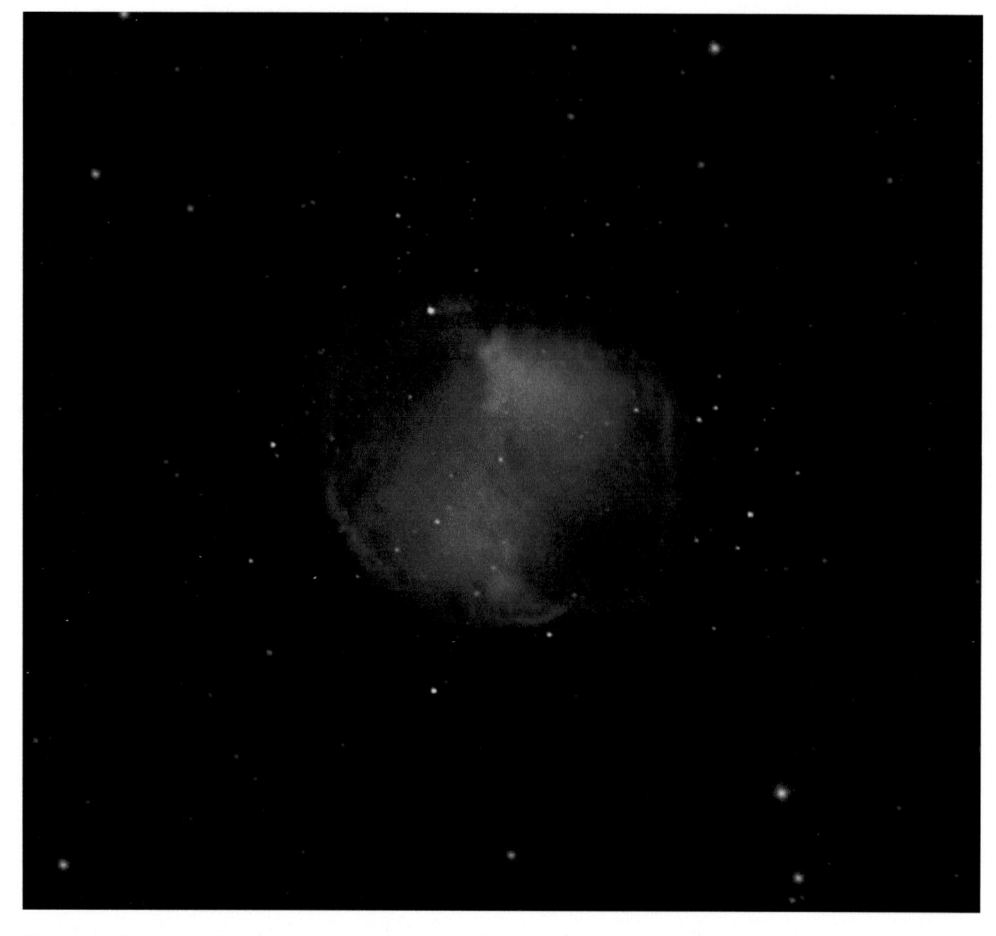

Figure 12.7 The final, cropped, image of the Dumbbell Nebula having applied two enhancement techniques

13
Imaging Planets with an Astronomical Video Camera or Canon DSLR

One way to try to mitigate the effects of turbulence in the atmosphere when taking planetary images is to take a video sequence of many, very short exposure, frames. Some of these, taken when calmer parts of the atmosphere lie along the line of sight, will be sharper than typical frames. Processing software such as *Registax* and *AutoStakkert! 2* can then analyse the frames to find the sharper images and then align and stack them to give a far better image than any single frame could provide. The technical name for this process is 'lucky imaging'.

Many astronomical video cameras (called 'webcams', as these were initially used by amateurs) are available, employing either colour or monochrome sensors with varying sensitivity and size. I would strongly advise beginners to first buy and use a colour camera – the imaging and subsequent processing is far simpler.

The software provided with the camera will take video sequences at a rate of 30–60 frames per second. A sturdy mount must be used with excellent tracking to keep the planetary image on the sensor during the taking of the video. A real problem is that of acquiring the image onto the few millimetre sized webcam sensors in the first place. The solution to avoiding much frustration is to employ a 'flip mirror' system, as shown in Figure 13.1. First the mirror is angled so that a relatively wide field image can be viewed with an eyepiece and the planet centred in the field. This could then be replaced with a narrow field eyepiece which could have illuminated crosshairs and can usually be made parfocal with the webcam sensor. Having accurately centred the planet, the mirror is then 'flipped' so that its image falls on the sensor.

Most webcams will come with their own capture software, such as *IC Capture.AS* which is used with the two Imaging Source cameras (one colour, one monochrome) that I have. However there is one freeware program (though a donation would be appreciated) called *FireCapture* (www.firecapture.de), which provides support for a very wide range of webcams and has a highly advanced user interface. This is now being used by many planetary and lunar imagers.

I believe that it is best for the tracking to be 'not quite perfect' to allow the planetary image to move *very slowly* across the sensor as a video sequence of one to two thousand frames is taken. This better samples the colour when a colour webcam is

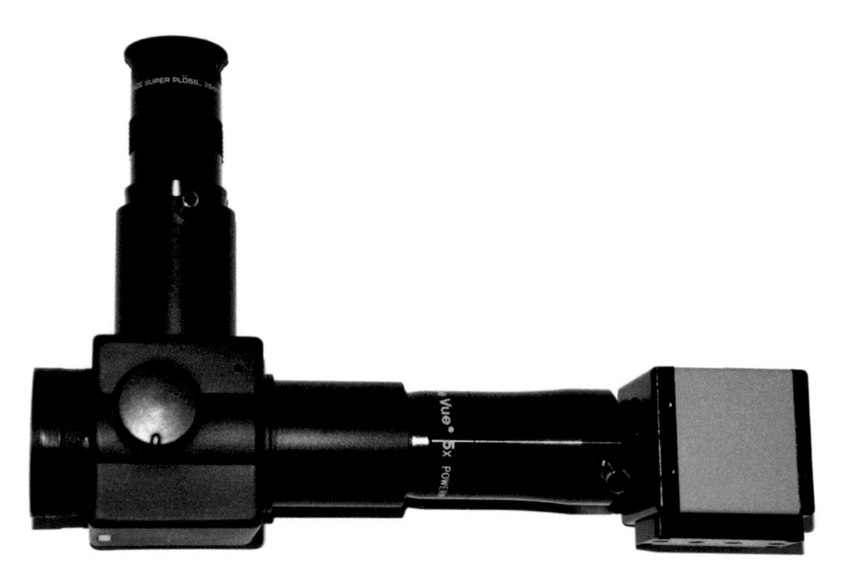

Figure 13.1 At left, the Vixen Flip Mirror with 26 mm eyepiece, centre, the TeleVue ×5 Powermate and, right, the Imaging Source DFK 21AU04.AS colour video camera used to acquire these images

used (as individual pixels are only sensitive to one colour) and also helps to remove the effect of any dust on the sensor. (When, inadvertently, the tracking of my mount was effectively perfect and the image stayed stationary on the webcam sensor, the pattern of the RGB Bayer matrix was faintly visible in the resulting image.)

Even with the majority of longer focal length telescopes, the image of the planet on the sensor will be rather small and so a Barlow lens is employed to increase it. But making the image larger will reduce its brightness, so each exposure will have to be longer and fewer sharp images will result. A compromise must be reached, and it turns out that under typical seeing conditions an effective focal ratio of f/20 is about right, so a ×2 Barlow would be used with a f/10 Schmidt–Cassegrain. If the seeing conditions are excellent, an effective focal ratio of f/30 to f/35 will provide better results when imaging Jupiter and Saturn and perhaps even more when imaging Mars.

I prefer to use the version 4 of *Registax* for planetary imaging as I feel there is more control over the process than in the later *Registax 6*. The following text and figures show how a 1392 frame video of Jupiter was processed. This had been taken with an Imaging Source DFK 21AU04.AS colour camera attached to a 127 mm, f/7 refractor. A TeleVue ×5 Barlow, giving an effective focal ratio of 35, was used as the seeing was excellent. The effective resolution was limited by the aperture of the telescope and the use of a larger telescope may well have given a better image.

Having loaded the video sequence into *Registax 4*, an alignment box of 256 pixels width is placed around one of the sharper images (found by moving the grey slider at the bottom to view the individual frames). A window then appears showing a

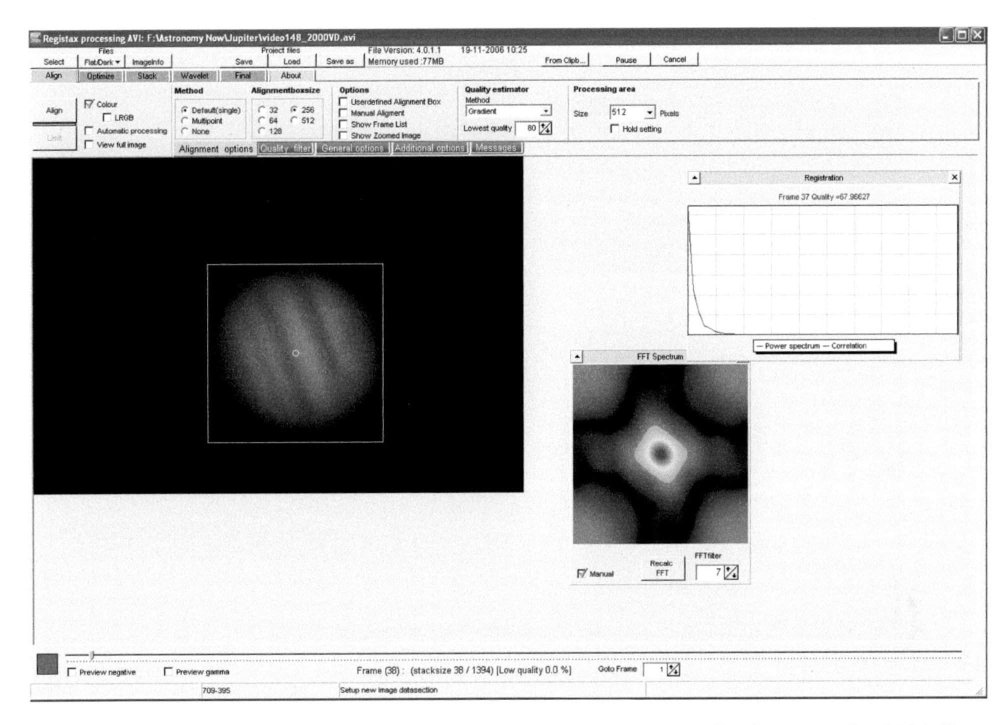

Figure 13.2 Putting an alignment box around the planet and adjusting the FFT filter parameter to remove any bright regions from the corners of the FFT display

Fourier transform of the planet. The FFT filter parameter should be increased until the corners of the box are free of colour, as seen in Figure 13.2.

When the 'Align' tab is clicked, one sees the alignment box follow the planetary image as it moves slowly across the sensor. When complete, the 'Limit' tab is clicked. The program has then ranked each frame in terms of image quality and has measured their X and Y positional offsets (in pixels) from the initially selected frame.

The next step is to produce a higher quality reference frame. By clicking on the 'Create a Reference Frame' tab, the program stacks some of the sharpest frames to produce a higher quality image. This is then used to refine the X and Y offsets found in the initial pass. I generally allow more than one optimising pass by not ticking the 'Single run optimiser' box before these are initiated by clicking on the large 'Optimise' tab at the left of the screen (Figure 13.3).

This complete, the small 'Stack' tab in the top row is clicked on and the 'Show Stack graph' box ticked. A window then shows all the frames ordered from left to right in rank order. The grey slider below the graph is then moved to the left to remove the less good frames from the stack and the slider at the top left of the window is moved down to remove those frames whose offset errors are more than most, as shown in Figure 13.4. This procedure selected 693 frames of the initial 1392 frames which were then stacked by clicking on the large 'Stack' tab.

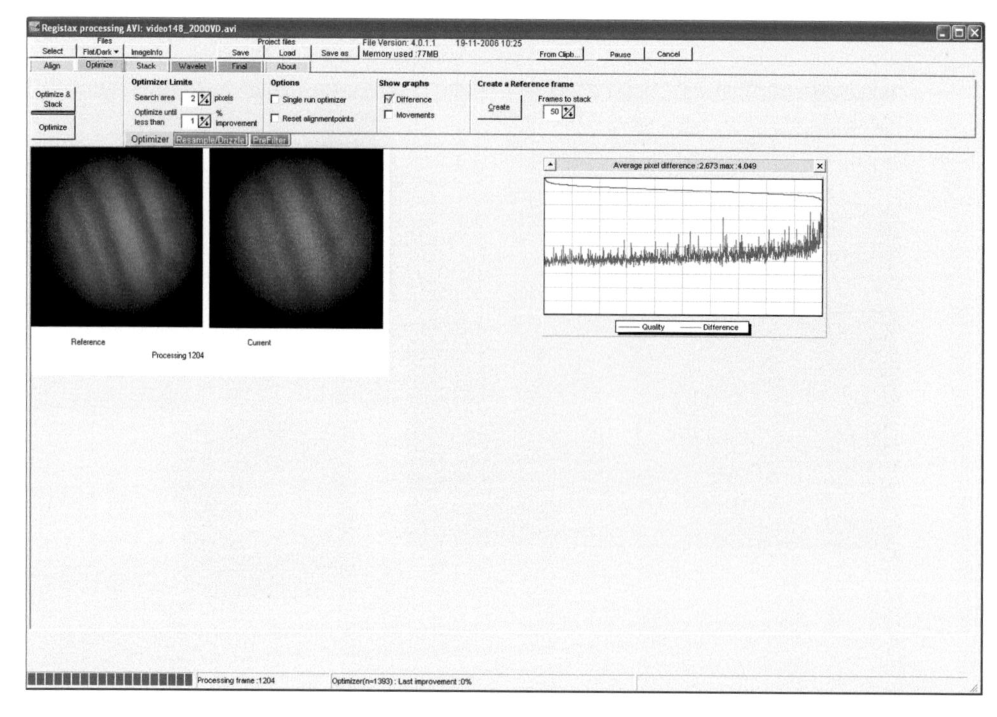

Figure 13.3 Optimising the positional offsets by comparing each frame (right) with the reference frame (left)

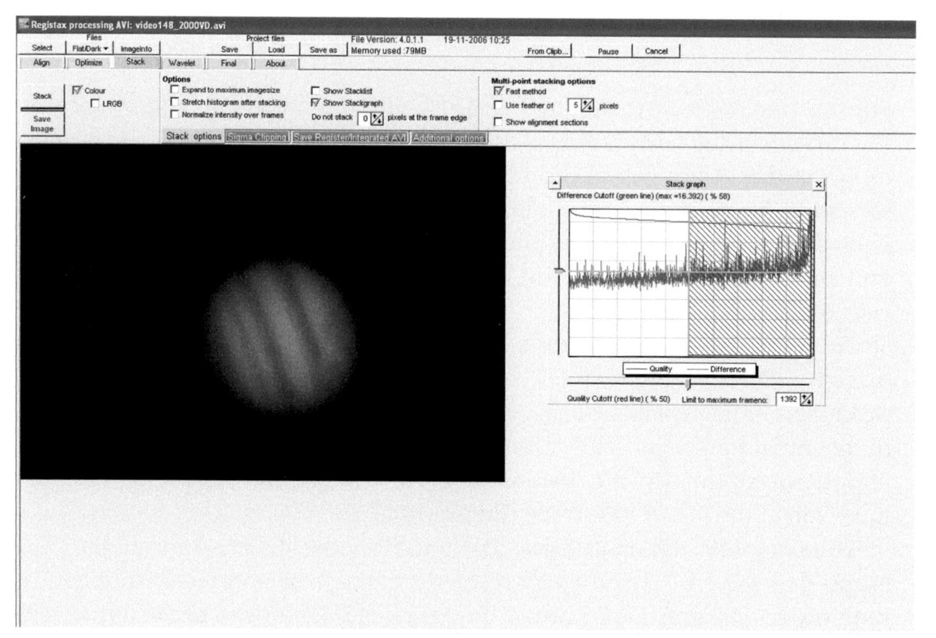

Figure 13.4 Using the Stack graph to limit the number of frames that will be stacked to provide the unsharpened image on left

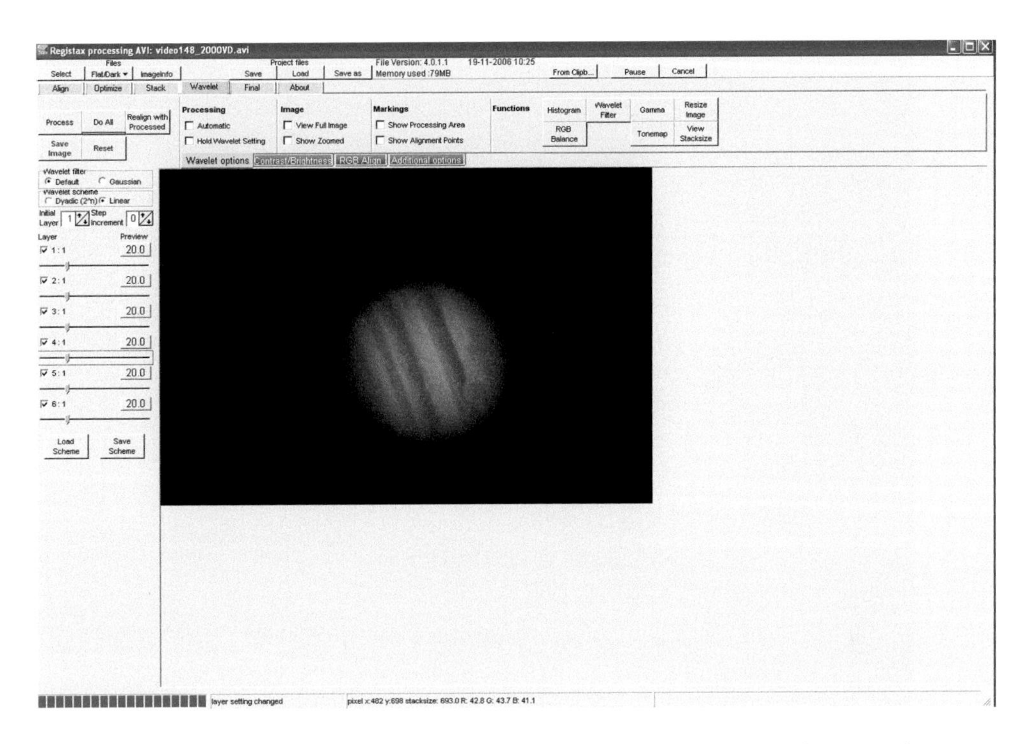

Figure 13.5 Using the wavelet sharpening window to provide a sharpened Jupiter image. In this case, all sliders were set to 20

The screen then shows the stacked image. At this point the 'Wavelets' tab is clicked and a set of six sliders appear in a window as seen in the Figure 13.5. By moving these sliders to the right, the image is sharpened. A good starting point would be with all values set to 10. The 'Do All' tab must then be clicked to apply the sharpening to the whole image before the resulting image is saved.

I suspect that many would find the resulting image, shown in Figure 13.6, somewhat disappointing when compared to the fantastic planetary images produced by Damian Peach (www.damianpeach.com), but his very best have been taken using a 14 inch telescope in Barbados, where the seeing conditions can be exceedingly good. The Jupiter image, produced on a night of good seeing, is pretty typical of what we can achieve here in the UK with a small telescope. Below is an image, processed in identical fashion, of Mars at its closest approach in 2014, when its angular size was just 15 arc seconds across. Given the telescope's resolution of one arc second, one would not expect to see great detail, but the image compares well with a theoretical image produced by the free program *WinJUPOS*, showing (at high resolution) what could be seen when the video sequence was taken.

Figure 13.6 Images of Jupiter, top, and Mars, left, taken using a 127 mm refractor and Imaging Source DFK 21AU04. AS colour video camera. The right hand Mars image is a simulated image produced by the free program, *WinJUPOS*

RGB Imaging

This requires the use of a mono camera such as the DMK 21AU618.AS or Point Grey Flea3 and a filter wheel to take the red, green and blue filters. Owing to the fast rotation rates of Jupiter and Saturn, care must be taken in producing the image. To achieve the full resolution that a large telescope (say 11 inches or more) is capable of for these planets, the time over which the three video sequences are taken should not exceed about three minutes, whilst for Mars 10 minutes is about the practical upper limit. It therefore helps if a camera capable of taking and downloading images faster than 30 frames per second is used to enable the best quality images to be produced. Some of the latest cameras are using the USB 3 interface protocol to help achieve speeds of up to 60 frames per second at full resolution. (If a smaller aperture telescope is to be used, then these times can be extended somewhat as the resolution will be limited by their aperture rather than the atmospheric seeing.)

There are two types of RGB filters. The cheapest filters use glass doped to absorb the unwanted two colours and are essentially the same as the RGB filters used in a DSLR Bayer matrix. Their passbands will be broad and will overlap somewhat. They do not incorporate an infrared cut-off filter, so one is needed in front of the CCD chip and is screwed into the camera's 1.25 inch barrel. Dichroic RGB filters that have much tighter passbands and greater transmissivity are available at higher cost and are well worth having for deep sky imaging, and can obviously be equally well used for planetary webcam imaging. They are discussed in detail in Chapter 21. As it is important to be able to switch between filters very quickly, a filter wheel is a necessity. Quite often the time spent on capturing the blue image will be need to be longer, and a good approach, advocated by Damian Peach, is to take the sequence R-G-B-G-R with the times used to take the R-G-B and B-G-R being within the limits mentioned above. It is then possible to produce two images by utilising the blue channel twice over.

Producing a Colour Image from the Red, Green and Blue Channels

Having processed the video sequences in a program such as *Registax* as described above, the red, green and blue images must then be combined to form a colour image. Each of the three images must be in 8-bit greyscale mode. At this point the relative brightness of the three images should be checked using the Levels box, Image > Adjustments > Levels. It may well be that the blue image is less bright than the other two, and if so, the right hand slider should be moved to the left to increase its brightness. It is very likely that the three planetary images will not be in precisely the same location within the frames, and if so, they can be combined and aligned using the following method.

First open a new image in RGB mode whose size is equal to or larger than the RGB images. If the 'Channels' window is then opened, three empty R, G and B channels will be seen. First open the Green mono image, select it (Ctrl-A), copy it (Ctrl-C) and then click on the appropriate channel box and paste the image into it (Ctrl-V). Now repeat with the red image. The result will look pretty awful as it is unlikely that the

Figure 13.7 Images of Jupiter and Saturn by Christopher Hill and Damian Peach

two will be aligned. The Move tool is then selected and by initially using the mouse and then, for fine adjustments, the cursor keys, the two can be accurately aligned. (This assumes that an equatorial mount has been used, so there is no frame rotation between the three RGB images. If not, two of the images may need to be rotated prior to combining them, as described in Chapter 21.) Finally add and align the blue image and the final colour image should appear.

In Figure 13.7, the Jupiter image by Christopher Hill was taken during the apparition of 2012 when Jupiter was at high elevation, whilst that by Damian Peach was taken back in 2003 when Saturn was far higher in the sky as seen from northern latitudes than in the apparition of 2013.

The Use of *WinJUPOS* to Align the RGB Images

A use of the *WinJUPOS* program is to give, for any time, a graphic of the face of a planet that would be presented to Earth, and I used it to give a reference image with which to compare my Mars image in Figure 13.6. But for planetary imagers there is a further use: *WinJUPOS* can be used to allow red, green and blue images to be combined even if they have been taken over a longer period than the few minutes referred to above. To achieve this, it projects each colour image onto a sphere, rotates each image to a common angle and projects it back onto a flat plane. The resulting three colour image can then be stacked. It is critical that the (UT) time of each video sequence is logged to an accuracy of a few seconds. A watch needs to be synchronised to UT and an excellent way of achieving this is to use the 'Official NIST US time' service: www.time.gov. This 'pings' one's computer to find the time delay in transferring data from the USA to one's location and then provides a display giving the precise UT. The date and the central time of each image are used to rename the file in the appropriate format. *WinJUPOS* then calculates the position, angular size and orientation of the images and saves the each data set in an 'IMS' file. The program can then derotate each image to a common angle and combine them to provide the final colour image of the planet. The process is covered in great detail in a PDF downloadable from www.sunspot51.com/Misc/winjupos.pdf.

Planetary Imaging with a DSLR Camera

Until I read an excellent article by Jerry Lodriguss in the magazine *Sky & Telescope* I had not even thought that this was sensible or even possible. His CD-based 'book' – *A Guide to a DSLR Planetary Imaging* – is available on the Internet[1] and is a superb introduction to planetary imaging, whether using a webcam or DSLR, and it includes excellent tutorials on using the alignment and stacking programs. Not only can many DSLRs be used for planetary imaging, they can do it well, and if you have a suitable camera, it would be well worth trying out the technique before purchasing a webcam.

The majority of new DSLRs have the ability to take video sequences and so it is not unreasonable that one could be used to undertake planetary imaging. However, just taking a video sequence as one would normally do will not work well! The basic problem is that each frame of the video will likely have an image size of 1280 × 720 pixels, but this image has been taken with a sensor that might have, as in the case of my Canon 1100D, a sensor size of 4272 × 2848 pixels. Either the image from a group of adjacent pixels (roughly 3–4 pixels wide) will be averaged to produce each output 'pixel' or a grid of pixels within the sensor will be used. In either case the AVI frames would provide a heavily under-sampled image of the planet.

Happily, there can be ways round this which are somewhat dependent on the camera. The easiest solution results if the camera is a Canon DSLR which has a Live View mode – as virtually all now do. You will, of course, also need a T-mount to attach the camera to the telescope. All that is then necessary is to download the free program called *EOS Camera Movie Record* (https://sourceforge.net/projects/eos-movrec/) and attach the camera to the computer with a USB cable. When the program is opened and the camera turned on, the Live View is seen on the computer screen with

[1] *A Guide to DSLR Planetary Imaging*, Jerry Lodriguss. Order from website: www.astropix.com/gdpi/gdpi.html

a white rectangle superposed in the centre. In the control panel is a box containing 5× beside the word zoom. If this is pressed, this part becomes the full frame so that all pixels in this area will be utilised, exactly as we want. A button on the extreme left enables a folder to be selected in which to save the video files and then, by clicking on the red 'Write' button, the capturing of the movie sequence is initiated. (Examples of the use of a Canon DSLR with *EOS Camera Movie Record* are given in Chapter 14 (see Figure 14.1) and Chapter 15 (see Figure 15.3) when used to produce high resolution images of the Moon and sunspot groups, respectively.) This program records files that can be immediately opened by *Registax 6* or *AutoStakkert! 2*. If *Registax 4* cannot read its files, a free program called *VirtualDub* (www.virtualdub.org) will convert them into standard AVI files.

A program called *Images Plus* (www.mlunsold.com/ilcameracontrol.html) will also capture Live View images from a Nikon DSLR, but it is rather expensive and it might actually be cheaper to purchase a second hand Canon body. Some cameras, the Canon 60D, have a 640 × 480 pixel 'Movie Crop Mode' that can be accessed from the video recording menu. This uses all the pixels from the central part of the sensor – just as required – at rates of up to 60 frames per second. In this case, the video file is recorded onto the SD card within the camera. (You will need a high capacity card!) These files are compressed and a program, such as the free program *SUPER* (http://super.en.softonic.com/), will be needed to convert these files into uncompressed AVI files for use by the stacking programs.

Processing Planetary Images in *AutoStakkert! 2*

This is a relatively new program (www.autostakkert.com) that is also free – although a donation would be appreciated – and which is now used by many of the world's leading planetary imagers. It currently lacks the ability to process most compressed AVI files, so a free program such as *VirtualDub* must be used to produce an uncompressed file first. Neither does it include a wavelets sharpening tool, but it does, on the other hand, provide two resulting images, one unsharpened and one sharpened. For planets, this latter image seems to provide a very good, not over sharpened, result.

The processing sequence is very similar to using *Registax*. The uncompressed AVI file is opened by clicking on the 'Open' tab in the control window. The first image of the sequence then appears in the frames window. It appears that the program automatically searches through the frames to find one of good quality. As in *Registax*, I prefer to use a single alignment box for planetary images, in which case the 'Single' button is clicked in the Alignment Points box and, using the mouse, a box is made to just surround the planet's disc. To estimate the image quality, a gradient method tends to work well with planetary images, so the 'Gradient' button in the Quality Estimator box is clicked.

The 'Analyse' tab is then clicked on and a progress bar is seen to move across until the process is completed, when a quality graph appears. The green bar can be moved across to view individual images that are ranked from left to right and can then be set so that a suitable number of frames will be stacked. Setting the 'cut-off' at the 50 per cent level seems a good compromise. In the 'Stack Options' box, I would click on the 'TIF' button (actual spelling on the screen) and click the 'Sharpened Images' box. The final tab to click on is 'Stack', when, again, a bar shows the progress in first aligning and then stacking the selected frames. When this is completed, two new sub-folders containing the resulting images appear in the folder in which the AVI file was selected.

Planetary Imaging at Low Elevations

Sadly, over the next decade, planetary imaging from the northern hemisphere will become more difficult as the major planets drop down towards the southerly part of the ecliptic – and so will be at low elevations even when crossing the meridian. The problem is that the atmosphere acts as a prism and so the planet as seen in different parts of the spectrum will be shifted in vertical position. Termed atmospheric dispersion, this blurs the planetary images and produces colour fringes at their limbs. The lower the elevation – and hence the greater the amount of atmosphere the planet is viewed through – the worse the effect. As an example, a stellar image will form a vertical spectrum that is 2.5 arcs seconds in angular size at an elevation of 30° but which increases to 4 arc seconds at an elevation of 20°. One can easily imagine the effects on an image of Mars, whose angular size was 18 arc seconds and at an elevation of ~18° when observed from the UK in May 2016.

This section will look at two ways that planetary imagers can mitigate the effects of atmospheric dispersion. First, let's consider what can be done if the video sequence of a planet is made with a colour webcam. Figure 13.8 is a screenshot of Jupiter taken when the planet was at a relatively low elevation and processed as usual in *Registax.* To improve the adjustment accuracy, the image has been exported, resized upwards by a factor of two and imported back into *Registax*. On the right hand side of the screen are a number of function tabs, one of which is called 'RGB Align'. When clicked upon, this opens up a window that allows the red and blue channels, of the image to be moved, pixel by pixel, with respect to the green channel so, as seen in the inset, the effects of the atmospheric dispersion can be significantly reduced.

When using a monochrome webcam and colour filters, the combining of the red, green and blue individual monochrome images as described above will perform a similar correction to mitigate the effects of atmospheric dispersion.

These two techniques cannot, of course, remove the effects of dispersion across the red, green and blue passbands, this being particularly significant in the blue, so the effects of the atmosphere can be only partially removed. There are, however, devices that can be placed in front of the imaging webcam that can achieve full correction. These, not surprisingly, are called 'Atmospheric Dispersion Correctors' (ADCs) and are made by a number of manufacturers, such as ZWO, Pierro Astro and Astro Systems Holland. They use two prisms to reverse the colour separation caused by the atmosphere. When the two are aligned as in Figure 13.9 (a), there is no dispersion

Figure 13.8 The use of *Registax* to align the red, green and blue channels to reduce atmospheric dispersion

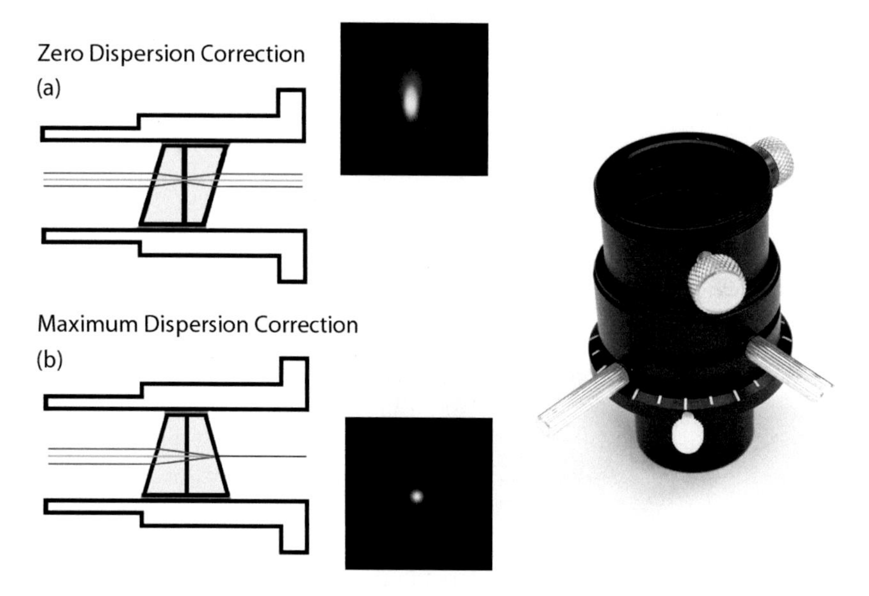

Figure 13.9 The dispersion correction is adjusted by rotating the two prisms using the control levers as seen in the image of the ZWO Atmospheric Dispersion Corrector

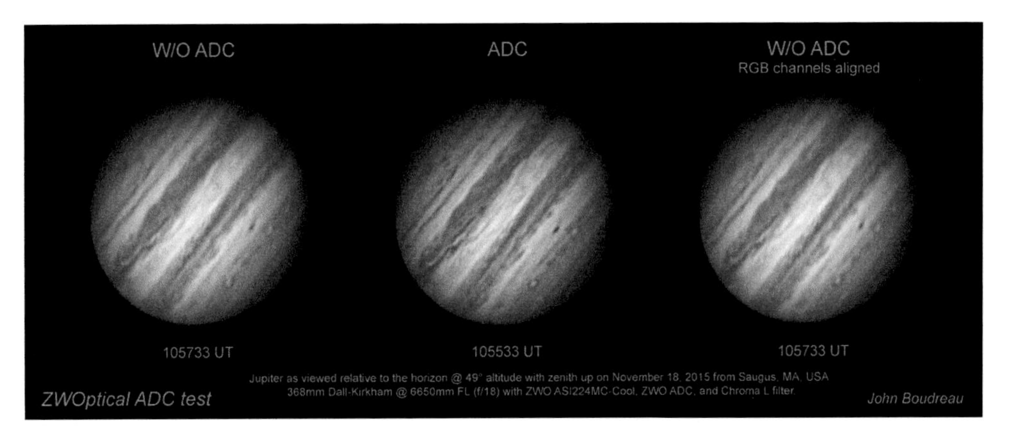

Figure 13.10 Jupiter imaged by top planetary imager John Boudreau: comparing an uncorrected image, left, with that corrected by the use of a ZWO Atmospheric Dispersion Corrector, centre, and where the red, green and blue channels have been aligned, right

correction applied, but when, as in Figure 13.9 (b), the two are rotated equally away from the null position, the amount of correction increases.

They work best at long effective focal lengths, so should be located between the Barlow lens (usually employed to increase the effective focal ratio) and the webcam. Atmospheric dispersion always spreads the light out in a vertical direction, so the ADC needs to be orientated so that its overall dispersion axis is also vertical. Usually, as in the case of my ZWO ADC, this is simply done by ensuring that the midpoint of the two prism adjustment levers is horizontal.

The prism angles must then be set to the appropriate rotation. To achieve this, the exposure of the planet's image seen in the video capture preview screen should be set so that, while the central part of the disc may be burnt out, the dimmer limb should show obvious red/blue tinges on opposite sides caused by the effects of atmospheric dispersion. To find the optimum setting of the prisms, the ADC levers are moved equally away from the zero point until the colour is seen to be even all round the limb. As an alternative, a star at the same elevation could be observed first: without correction a vertical spectrum will be seen and the prisms are adjusted until a clean star image is produced.

Well worth reading is an excellent detailed discussion of Atmospheric Dispersion Correction that can be found at http://skyinspector.co.uk/atm-dispersion-corrector--adc.

14 Video Imaging of the Moon with a Webcam or DSLR

As described in the previous chapter, both Canon cameras (using the program *EOS Movie Record*) and webcams can be used to take video sequences which are then processed in programs such as *Registax* and *AutoStakkert! 2*. Exactly the same techniques as used when imaging the planets can be used to image the Moon, so helping to remove the effects of atmospheric turbulence and thus provide sharper images.

In contrast to the 23 × 15 mm sensor size of a typical DSLR, the sensor size in a typical webcam is just 5 × 4 mm. This means that an individual image will cover a relatively small area of the Moon's surface. The result is that one needs to take quite a number of images and then compose them into one final image. By hand this would be an incredibly time consuming job but, as mentioned above, there are programs that will do it for you. I am now using a free program called *Microsoft ICE* which, in this application, is well-nigh miraculous. One needs to make sure that there is plenty of overlap between the individual frames that will make up the final image – for one particular lunar image, I made 22 individual frames when ~14 would have just covered the Moon. The area of the Moon covered by an individual image will obviously depend on the telescope's focal length. The numbers that I have just given relate to my Taskahashi FS102, 800 mm focal length refractor. If I were using my 2350 mm focal length Schmidt–Cassegrain, the images would cover about one ninth of the area and it would require well over 100 individual images (called panes) to be processed!

As a simple example, using a modified Canon 1100D and short focal length refractor, I imaged the 7.3-day old Moon. In this case, four video sequences were needed to cover the disc, allowing suitable overlaps between them. Figure 14.1 shows the *EOS Movie Record* window before and after the ×5 mode was initiated. (Not all the program window appeared on the widescreen laptop display.)

Each of the frames in these video sequences then needs to be analysed to find and then align and stack the sharpest frames within the sequence. *Registax 6* was chosen to achieve this. The video sequence captured by *EOS Movie Record* is opened and the first frame is displayed. By moving the slider below the image, individual frames can be observed and, if it is obvious, one of the sharper ones selected. The effects of the

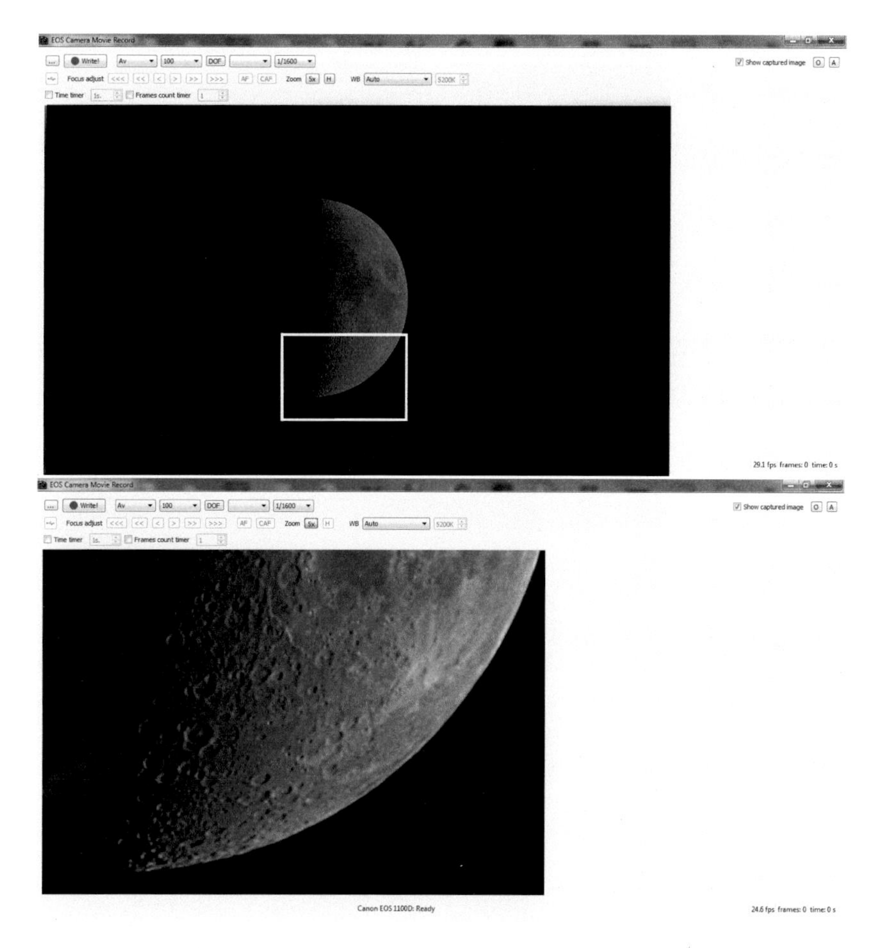

Figure 14.1 The *EOS Movie Record* windows on normal (top) and ×5 magnification (bottom)

seeing can be seen by comparing the two frames of a video sequence covering *Mare Nectaris* in Figure 14.2. The one on the left is significantly sharper. It is these sharper frames that will be selected to stack in *Registax* to produce the final image.

The effects of the seeing may differ across the image and so *Registax 6* allows a number of align points to be set within the frame. 'Set Align Points' is initiated with an align box size of ~30 pixels, when, in my view, far too many align points appear (Figure 14.3, top). By moving the associated slider across to the right the number is reduced and I tend to remove almost all of them and then, using the mouse, click upon obvious features such as small craters or mountain peaks to give around 30 align points (Figure 14.3, bottom).

The align sequence is initiated and, when complete, the 'Limit' tab is clicked upon. As shown in Figure 14.4, each align point then has a green line attached. Ideally, these should be short, meaning that the telescope tracking is good, and all in the

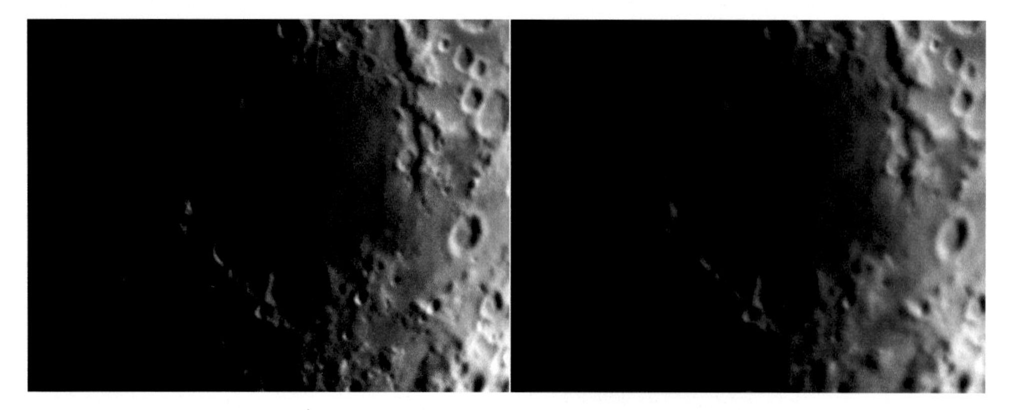

Figure 14.2 Two frames of a video sequence covering *Mare Nectaris* showing a significant difference in sharpness

same direction, meaning that the seeing is good. (This example was excellent.) If the 'Stackgraph' box is ticked, a graph appears showing the relative quality of all the frames, with the best on the left of the graph and the worst on the right. An associated slider can be moved to the left so that only the better ones are selected. This is a matter of judgement; fewer frames will give a sharper, but noisier, final image. There is a second vertical slider on the left of the stackgraph and this can be slid down to remove those frames where the movement of the image relative to the reference frame is large. Ideally, if the seeing is good, there will be a very gentle fall off in quality and the best (perhaps 90 per cent or better) are then stacked to produce the image that can be saved for further processing.

The *Registax* program does include a very powerful sharpening feature called 'Wavelets'. When initiated, a set of six sliders appears, and if each of these is moved gently to the right, one can observe the sharpening effect on the image. Giving each slider a value of ~10 could be a first setting, as shown in Figure 14.5. When a good (but not over sharpened) image results, the 'Do All' tab is clicked and the final result saved.

Having processed each pane identically in *Registax 6* and cropping each image to remove any edge effects, *Microsoft ICE* can then be used to combine the resulting panes into an image of the entire disc of the Moon simply by dragging the panes into its workspace. Figure 14.6 shows the *ICE* screen having combined the four panes, with the final image alongside.

Enhancing the Lunar Image

The image may well look rather flat with somewhat low contrast, particularly if a reflecting telescope has been used to make the image. It is surprising what an improvement can be achieved with a just few procedures. One procedure is simply to use the 'Levels' command in *Photoshop*. The first thing is to check the histogram in the (Image Adjust > Levels) box. Hopefully the initial image was not overexposed and so there will be a gap to the right of the histogram. It is best to leave this for the time being, but,

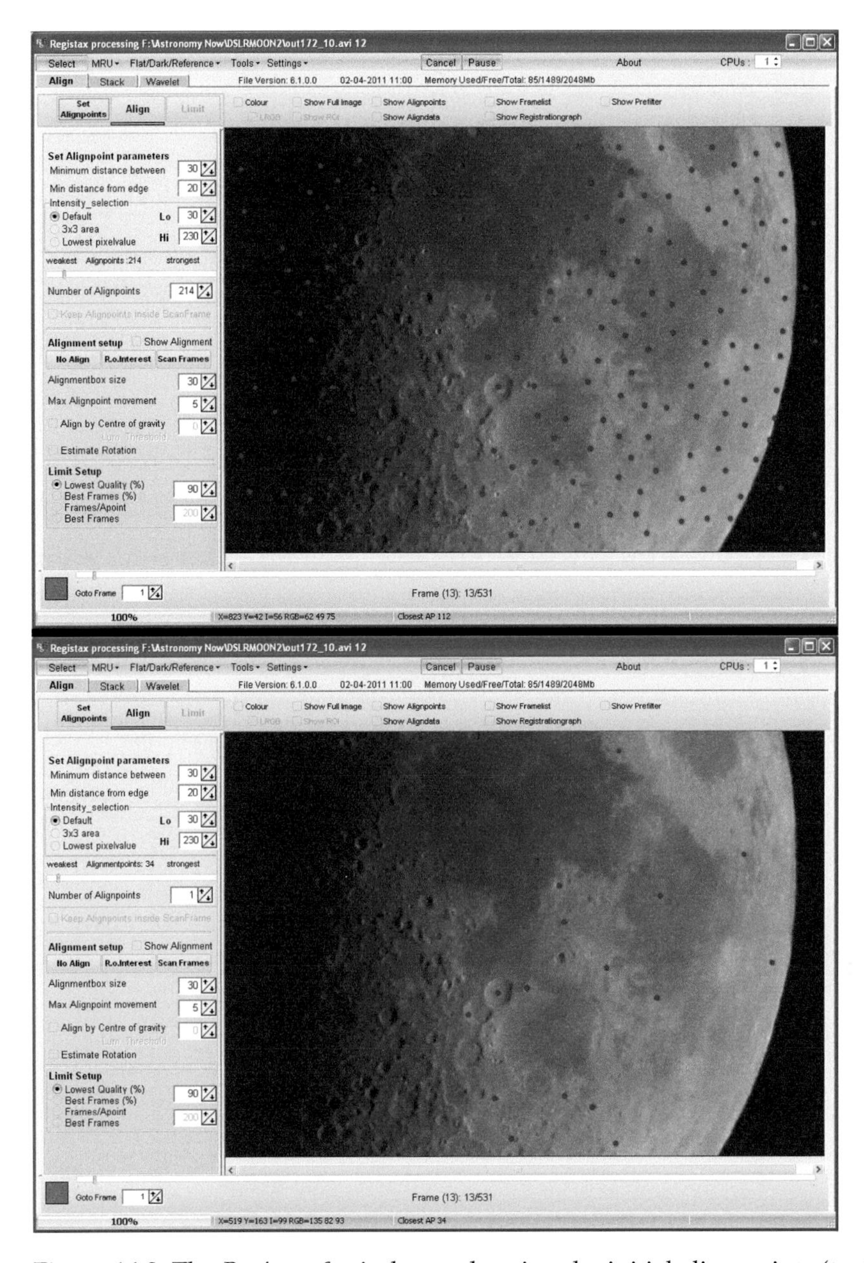

Figure 14.3 The *Registax 6* windows, showing the initial align points (top) and those manually selected (bottom)

with the preview box ticked, it may be worth adjusting the left and middle sliders to improve the image. However, it is important to be careful: if the left slider is moved too far to the right, detail on the limb of the Moon will begin to disappear.

Another procedure that is less obvious and not well known can make an amazing difference. For this to be used, there should be a gap between the right hand side

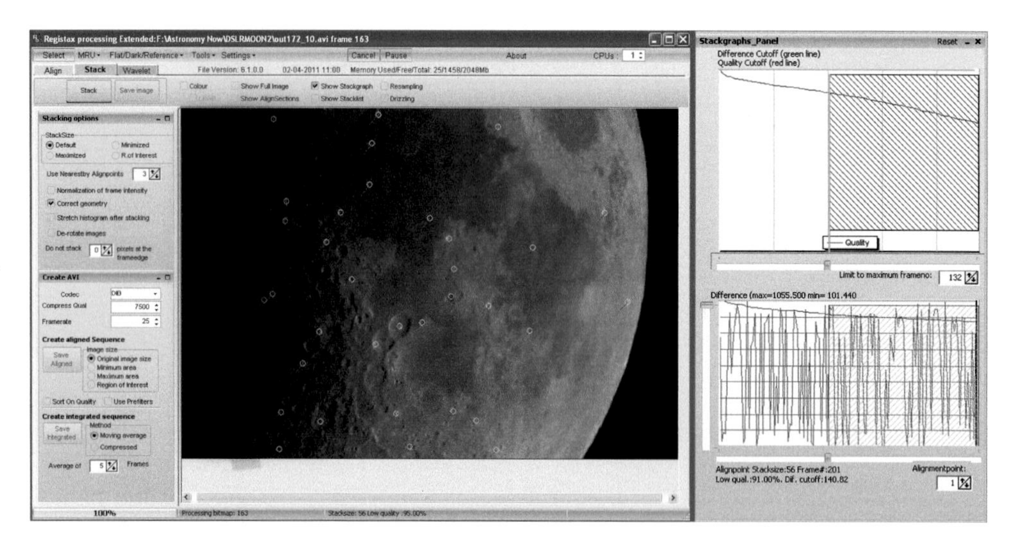

Figure 14.4 The *Registax 6* window having aligned the video sequence, showing excellent tracking and atmospheric seeing

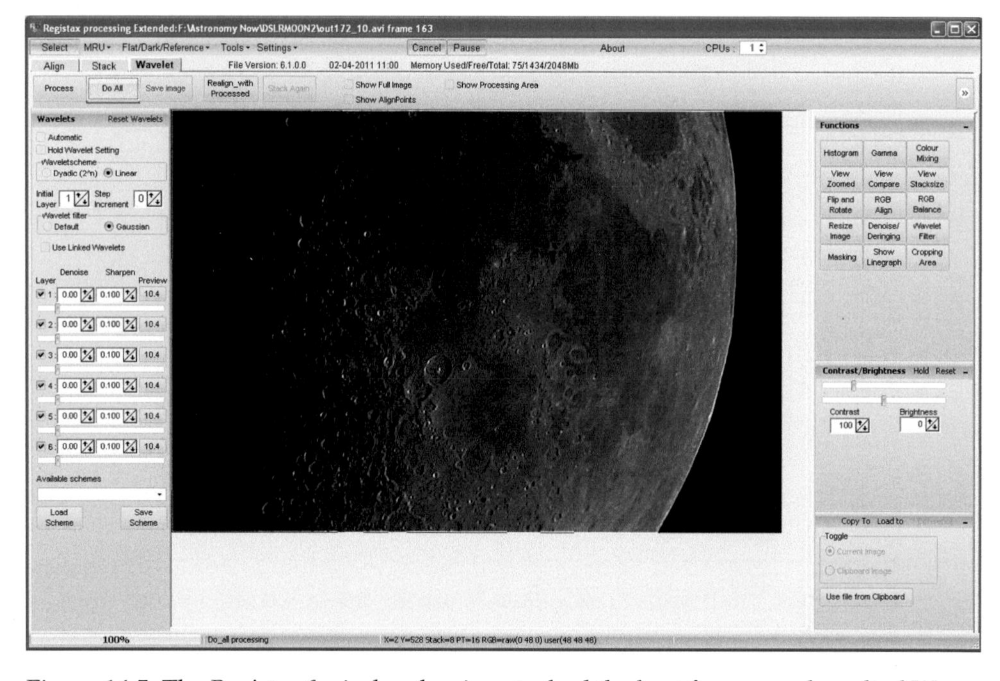

Figure 14.5 The *Registax 6* window having stacked the best frames and applied Wavelets sharpening to provide one of the four required panes to cover the Moon's disc

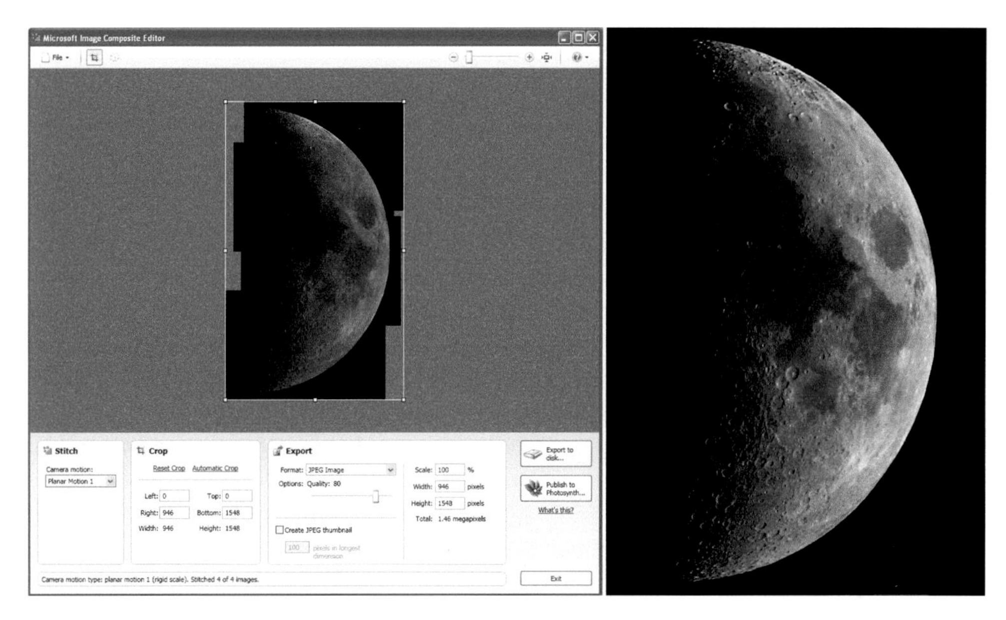

Figure 14.6 The *Microsoft ICE* window (left) having combined the four panes, with the final enhanced image of the 7.3-day old Moon (right)

of the histogram and the peak white (so the initial images must be a little underexposed) as the process tends to increase the brightness of the brighter parts of the image. Select the 'Unsharp Mask' filter (Filter > Sharpen > Unsharp Mask) and in the box move the radius slider right over to the right (it may then be set to 250 pixels). Then bring the amount slider down to between 10 and 30 per cent and observe the result with the preview box ticked. Adjust the slider to give an enhanced, but still natural, image. I think you will be impressed. One can then employ some gentle sharpening to give the final image.

Using a Webcam to Image the Full Moon and Jupiter

In November 2012, Jupiter came quite close to a full Moon and I felt that it would be nice to image the two together in the sky. At the time I was using an Imaging Source colour webcam type DFK 21AU04.AS with a 640 × 480 pixel array. It comes with a 1.25″ eyepiece adapter and image acquisition software. When connected through a USB port on my laptop, the software provides a liveview display to enable easy focusing of the image. Within the device properties, one can select either 'Auto Exposure' or set a specified exposure. For this particular imaging exercise I was recording compressed AVIs to minimise the amount of memory storage required on the controlling laptop's hard disk.

Moving over the Moon's surface, I selected an exposure of 1/1500 second so as not to overexpose the brightest regions of the surface. (This is quite important, so one should do an initial scan of the whole surface before selecting a suitable exposure.) The Takahashi FS102 was mounted on my iOptron Minitower alt-az mount. This is not ideal because of 'frame rotation' – that later images will be rotated somewhat relative to that taken first – however, *Microsoft ICE* appears to be able to correct for this. The camera was imaging 30 frames per second, so each ~1000 frame sequence took about 33 seconds. The telescope was simply driven over each part of the lunar surface and an AVI sequence taken, ensuring that there would be plenty of overlap so that *Microsoft ICE* could compose them into a final image. Having completed covering the Moon, the telescope was moved over to Jupiter and a single 1000 frame AVI sequence was taken with Jupiter suitably exposed. The Jovian moons were too faint to appear in this image, so the exposure was increased so that they were visible (but with Jupiter heavily overexposed) and a final AVI was taken. To cover the lunar disc, 14 video sequences were taken and, as described above, *Registax 6* was used to process them. *Microsoft ICE* was then opened and the 14 individual images selected as a group and 'dragged' onto its working area to compose the images into one. *Microsoft ICE* had distorted the overall image somewhat in the fitting process, so a compensating adjustment was applied in *Photoshop* using the 'Transform' tool.

The video sequences of Jupiter were processed in *Registax 4* as described in the previous chapter, the canvas size of the lunar image was increased (with a black background) and the Jovian disc and satellite images were overlaid, giving two additional layers. The satellite image was aligned with the Jovian disc image using the 'Move' tool and the image was flattened using the 'Lighten' blending mode to give the result shown in Figure 14.7.

Using a Monochrome Webcam to Image the Moon

The less expensive webcams have a quite small sensor, typically 5 × 4 mm in size. In some ways this is good as there is less data to transfer to the computer for each frame, which allows frame rates of up to 60 frames per second to be achieved. The small sensor does mean, however, that quite a number of individual video sequences will be required to cover the entire lunar surface. One of the best with which to image the Moon is the monochrome Imaging Source camera DMK 21AU618.AS, which uses a USB 2 interface to transfer up to 60 frames per second. It employs a very sensitive Sony ICX618ALA sensor, which has 5.6 micron square pixels in a 640 × 480 grid. One slightly unusual feature is that its peak sensitivity is in the red (having a quantum efficiency of 62 per cent) and the sensor still has a quantum efficiency of 50 per cent in the near infrared. This means that it is an ideal camera to image the Moon in the red or infrared part of the spectrum.

Imaging the Moon in the Red or Infrared

There are two reasons why one might wish to restrict the bandwidth used to image the Moon to the long wavelength parts of the spectrum. First, owing to scattering in the atmosphere and refraction (particularly if the Moon is low in the sky), the blue end of the visible spectrum will give a less sharp image than one taken at the red end of

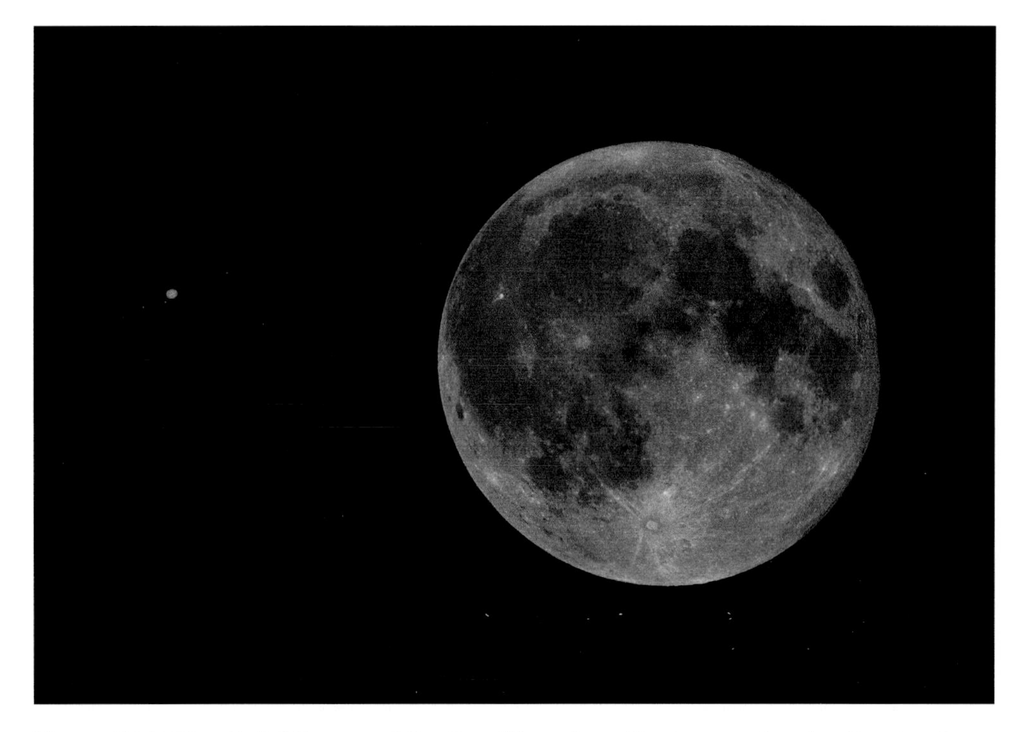

Figure 14.7 The Full Moon and Jupiter (To reduce the image size, Jupiter has been brought closer to the Moon than was, in fact, the case)

the spectrum. Secondly, the effects of atmospheric turbulence – causing the Moon to appear to be boiling at high magnifications – is far more disruptive to blue or green light than to red light and it affects the near infrared (700 to 1000 nanometres) even less. So it can be well worth restricting the wavelengths with which to image the Moon to either the red or infrared part of the spectrum. A dichroic red filter such as those made by Baader or Astronomik is best used to image in the red or the Astronomik IR 742 infrared filter (which has a 742 to 1100 nanometres passband at greater than 95 per cent transmissivity) used to image in the infrared. Lunar features show more tonal variation and lunar ray systems are more distinct when a red filter is used, but if the seeing is not excellent, an infrared filter might give a better overall result.

Image Scale

The lunar surface must be sampled by the pixels within the sensor so that all the available resolution that can be achieved by the telescope in use is captured. As the resolution possible is determined by the aperture of the telescope, larger aperture telescopes have the potential to provide higher resolution images when the seeing is exceptional. As an example, on a night when sadly this was not the case, I chose to image the Moon in the infrared using the DMK 21AU618.AS. When used with my

Vixen VC200L, 1.8-metre focal length telescope, each pixel of the camera is 'seeing' an area of the Moon ~0.6 arc seconds across. The Nyquist theorem requires that to extract all of the available information within the image, this must be a half or preferably a third of the resolution that one is hoping to capture. Thus, using this camera and telescope, the resolution of the resultant lunar image could be no more than ~1.2 arc seconds. Using a ×2 or ×2.5 Barlow lens in front of the camera would give a higher possible resolution should the seeing be excellent. There is a problem though, in that to cover even the 5-day old Moon's surface that I aimed to image using a ×2.5 Barlow would require well over a hundred individual panes, and even without using the Barlow, I took 45 video sequences (having plenty of overlap) to cover the lunar surface and so allow *Microsoft ICE* to combine them into an image. As the seeing was poor, I had decided to image in the infrared and this required an exposure of 1/125th of a second for each frame, which *IC Capture* recorded at a rate of 60 frames per second. Video sequences of ~2000 frames were captured and processed in *Registax 6*. The best ~400 frames of each were selected for stacking, the results of which were sharpened using the 'Wavelets' tool with all sliders at position 30.

The resulting 16-bit TIFF images were imported into *Photoshop* and very slightly cropped to remove any edge effects due to slight motion of the image across the sensor owing to turbulence and any residual tracking errors of the mount. These 45 TIFF panes were then selected as a group and dropped into *Microsoft ICE,* which quickly produced the entire lunar image as seen in Figure 14.8 (left).

The composite image was sharpened a little using the 'Smart Sharpen' tool and the local contrast increased slightly using the 'Unsharp Mask' filter with a large radius (88 pixels) and small amount (13 per cent). As seen in Figure 14.8 (left), the overall result was somewhat 'bland' compared to lunar images taken in white light. To 'lift' the image, I selected the *Mare* regions as seen in Figure 14.8 (right) and applied further local contrast enhancement using the 'Unsharp Mask' filter with a radius of 88 pixels and an amount of 22 per cent. This greatly helped to bring out their texture and features and significantly improved the overall image. Figure 14.9 shows the final image with two insets to enable the achieved image quality to be more easily seen. Owing, I think, to the fact that the Moon was imaged in the infrared, the image does have a rather different 'feel' to it than those taken in white light.

The resolution achieved was of the order of 1.5 arc seconds, limited partly by the resolution of the telescope (~0.7 arc seconds) but mostly by the relatively poor seeing. On the rare nights of exceptional seeing and using a telescope of 12 to 14 inches aperture, a resolution of 0.5 arc second is achievable. With this resolution, imaging the whole lunar surface is quite a challenge, but an interesting project is to individually image interesting regions of the lunar surface. An excellent list compiled by Charles Wood can be found at www.astrospider.com/Lunar100list.htm.

What is certainly the best lunar image ever taken by this method (using specialised webcams rather than DSLRs) can be found from this link: www.lunarworldrecord.org. A team of 11 top astrophotographers took close to 1000 panes, derived from 1.2 million individual frames, which were then combined together to give the final image. The image quality is stunning.

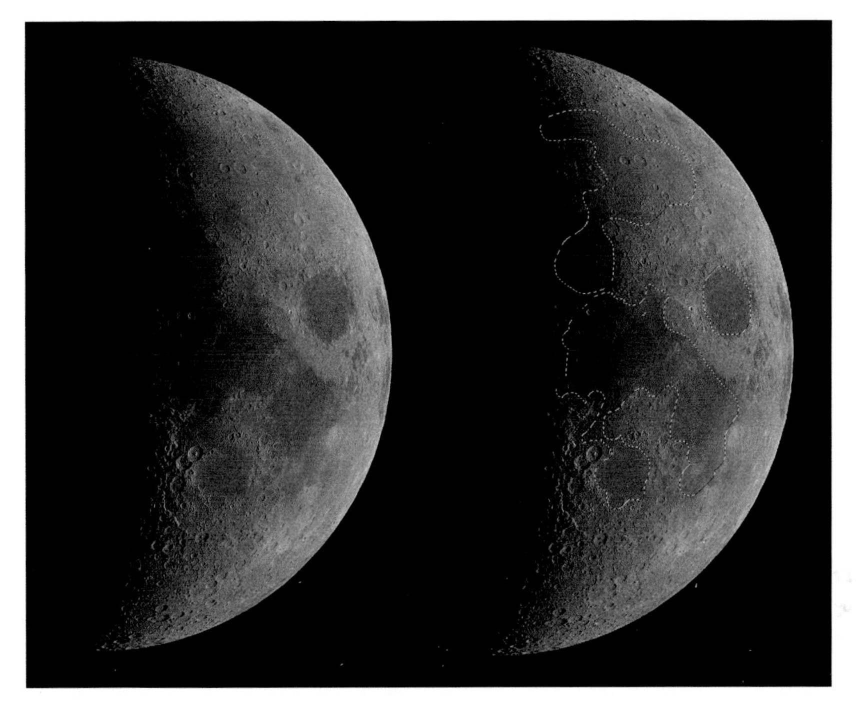

Figure 14.8 The initial infrared image (left) and the selection of the *Mare* regions (right) so that they could be enhanced

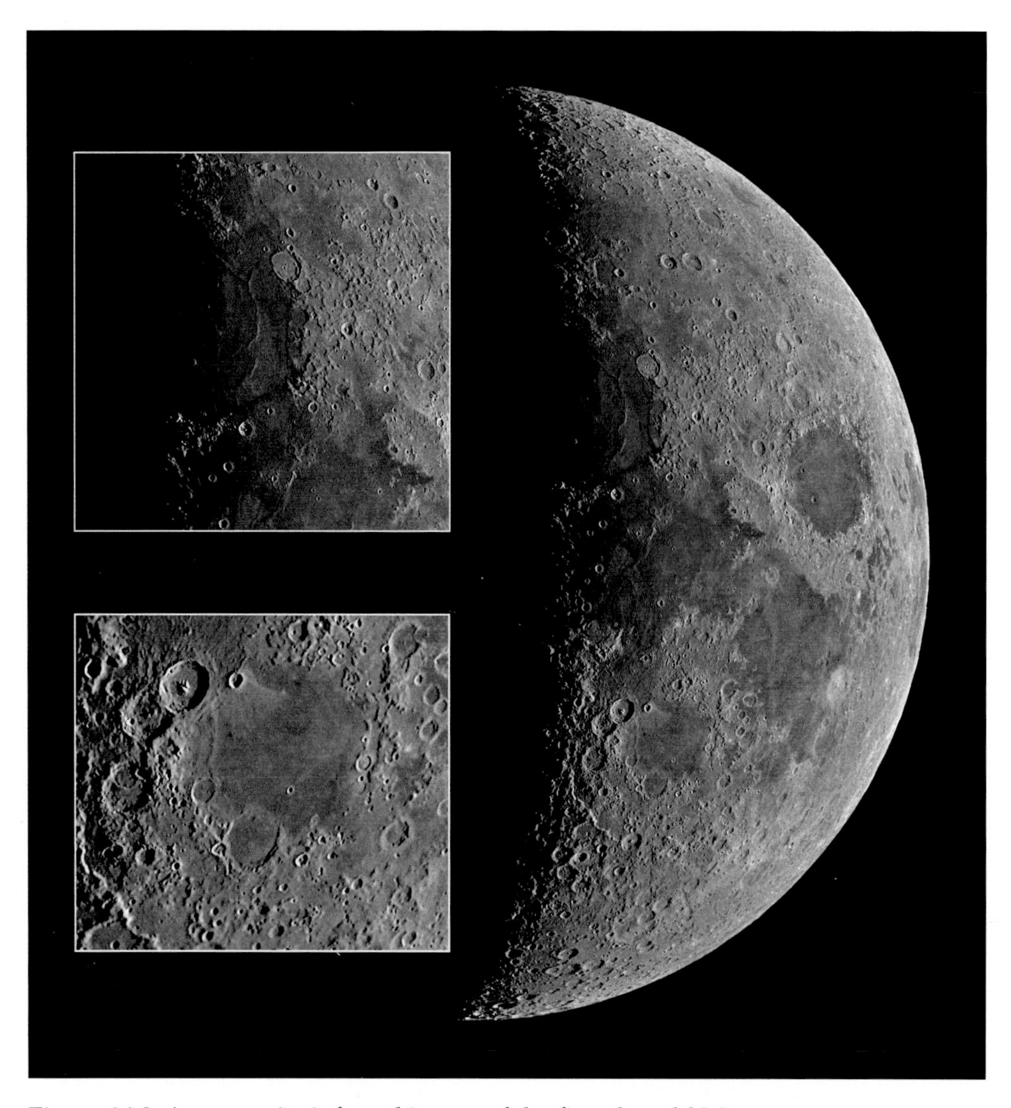

Figure 14.9 A composite infrared image of the five-day old Moon

15
Imaging the Sun in White Light

The Sun is the only astronomical object that could cause harm to an observer, so it is vital to follow the safety instructions provided by the manufacturers of the solar filters or Herschel wedges used to reduce the Sun's brightness to safe levels. In particular, a finder scope should be covered or removed or, even better, replaced with a KAYEM solar finder (shown in Figure 15.5), which can be directly obtained at low cost from www.rakm.co.uk. Without a solar finder, rather than try to align a telescope by eye, which implies looking up towards the Sun, a piece of white card can be held up behind the telescope and its pointing adjusted so that the telescope's shadow is circular.

Several manufacturers provide solar filters in a range of sizes to fit *in front* of the telescope aperture. Many use a silver metallised film made by Baader Planetarium or black polymer sheet made by Thousand Oaks Optical, while Seymour Solar can provide glass mounted filters. Those based on polymer sheets give a more natural colour to the Sun. For around £20 one can purchase both the Baader and the Thousand Oaks filter material in A4 sized sheets, which, given some 'Blue Peter' skills, can be made into suitable filters. It is vital that they cannot become detached from the telescope as the sensor of the camera (or your retina if observing visually) would be severely damaged. Figure 15.1 shows both types of filter in aluminium mountings.

There is an alternative method of reducing the Sun's brightness based on the use of a 'Herschel wedge'. *These can only be used with refractors whose objective sizes are no greater than 150 mm.* They look rather like a star diagonal and include a prism that reflects ~5 per cent of the Sun's energy towards the eyepiece or camera, with the remaining ~95 per cent absorbed in a heat sink at the back of the wedge. This is still far too bright and a neutral density filter is incorporated within the housing to reduce Sun's energy to safe levels. Further neutral density filters or a polarising filter (as the light reflected by the prism is polarised) can be used to adjust the brightness. It is also possible to incorporate a Baader Planetarium Solar Continuum filter, which cuts out all light except that limited narrow spectral region around 540nm, in the green part of the spectrum. When observing at this wavelength the granulation and the structure of the sunspots are revealed with the highest contrast. When imaging without a Solar Continuum filter, the contrast can be enhanced further by using an infrared blocking filter in front of the camera.

Figure 15.1 Baader Solar Film (silver) and polymer film filters and the Lunt and Baader Herschel wedges used to give safe brightness levels for observing or imaging the Sun

Two types are widely available: those sold by Lunt Solar Systems, available in 1.25 and 2 inch barrel versions, and the 2 inch barrel version manufactured by Baader Planetarium. As shown in Figure 15.1, the latter has a ceramic plate at its rear on which a diffused image of the Sun is seen. It can thus be used as the solar finder. It is generally agreed that Herschel wedge systems give the highest quality solar images, with filters using Baader Solar Film second best. A warning: some refractors may have insufficient inward focuser travel to enable a DSLR to be focused if a star diagonal or Herschel wedge is in the optical path, so it is worth first checking with a star diagonal that focus can be easily reached if one is thinking about purchasing a Herschel wedge. Baader produce a #27 adapter for use with its Herschel wedge to directly mount the DSLR onto the body of the wedge, so possibly overcoming this problem. (With some engineering expertise, it may well be possible to reduce the length of the telescope tube to allow focus to be reached, but then a 2 inch extender may sometimes be needed to be used for visual observing if the focuser extension is not too great.)

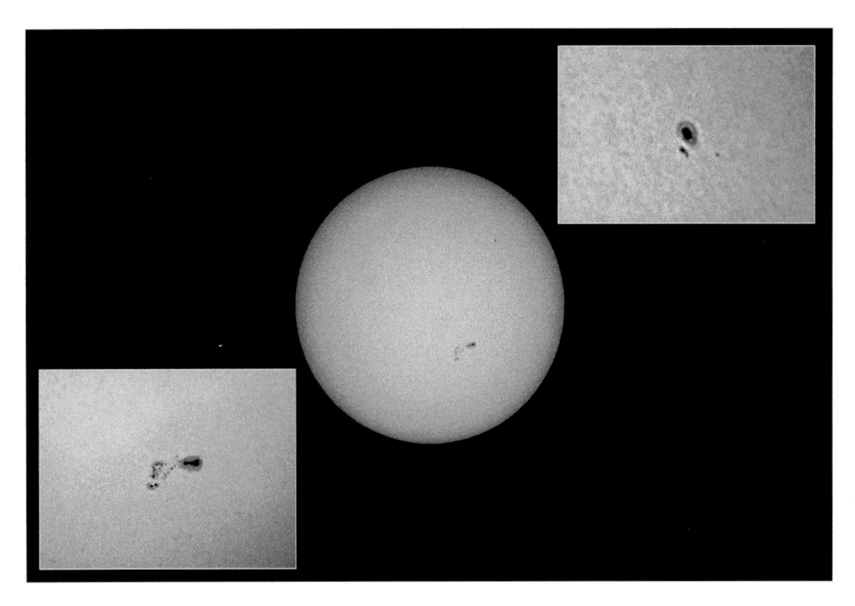

Figure 15.2 Full solar disc imaged with a 127 mm refractor, Baader solar filter and Canon EOS 1100D camera. The two insets show the sunspot groups imaged with the addition of a ×2 Barlow lens

Having achieved a solar image of suitable brightness, the method of imaging is exactly the same as when imaging the Moon as has been described in previous chapters. Either a DSLR can be used as normal to image the whole disc or, particularly to get higher resolution images of sunspot regions, a Canon DSLR having 'Live View' can be used as a 'webcam' using the *EOS Movie Record* program to produce video files. A webcam such as the Imaging Source DMK 21AU618.AS can also be used. Being monochrome, it will have higher sensitivity than a DSLR and its exposures will thus be less so, making it more likely that the very brief moment of exceptional clarity can be captured. In either case, a video sequence of perhaps 1000 frames is analysed in *Registax* or *AutoStakkert! 2*: the individual frames are aligned and the sharper ones stacked and sharpened to give the final image.

To illustrate this chapter, I have carried out some solar imaging exercises. Using a full aperture Baader Solar Film filter mounted on a 127 mm refractor, a Canon EOS 1100D camera was first used to image the whole disc on 8 August 2015. Using *Adobe Photoshop*, the image was converted into a monochrome image and then converted back to RGB colour mode. (This produces a monochrome image that is capable of being coloured.) This image was duplicated to give a second layer of the correct dimensions. The duplicate layer was coloured (covering over the duplicated image) using the paintbrush with an RGB colour mix of 253, 196 and 93 (as used by the SOHO[1] satellite). The final image, shown in Figure 15.2, was obtained by using the 'Colour'

[1] Solar and Heliospheric Observatory

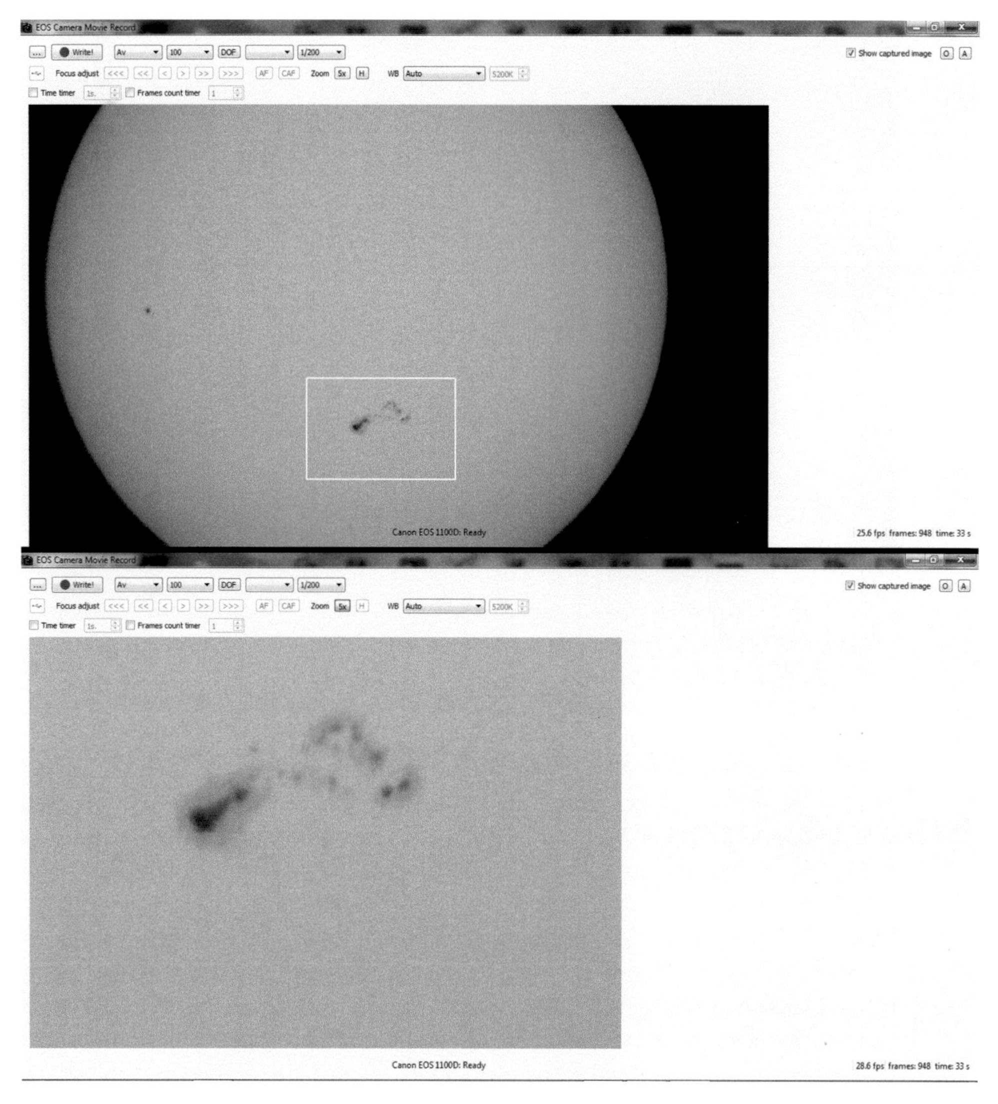

Figure 15.3 The *EOS Movie Record* screens at full size and ×5 magnification when coupled to a Canon 1100D camera whose 'Live View' was being captured and then saved as a video file

blending mode when the two layers were flattened. The insets show the regions around the two sunspot groups when a ×2 Barlow lens was placed prior to the camera.

The Sun's surface showed one interesting sunspot group (no. 2396) and this was imaged using *EOS Movie Record* using 2-inch ×2 and ×4 Barlow lenses prior to the camera. (As seen in Figure 15.3, the images appear pink, as the camera has been modified to be more sensitive to H-alpha emission.) The videos of ~1000 frames were aligned and the higher quality frames stacked in *Registax 6*. A small amount of Wavelet sharpening (all values ~10) was applied to the resulting images, which were then given some

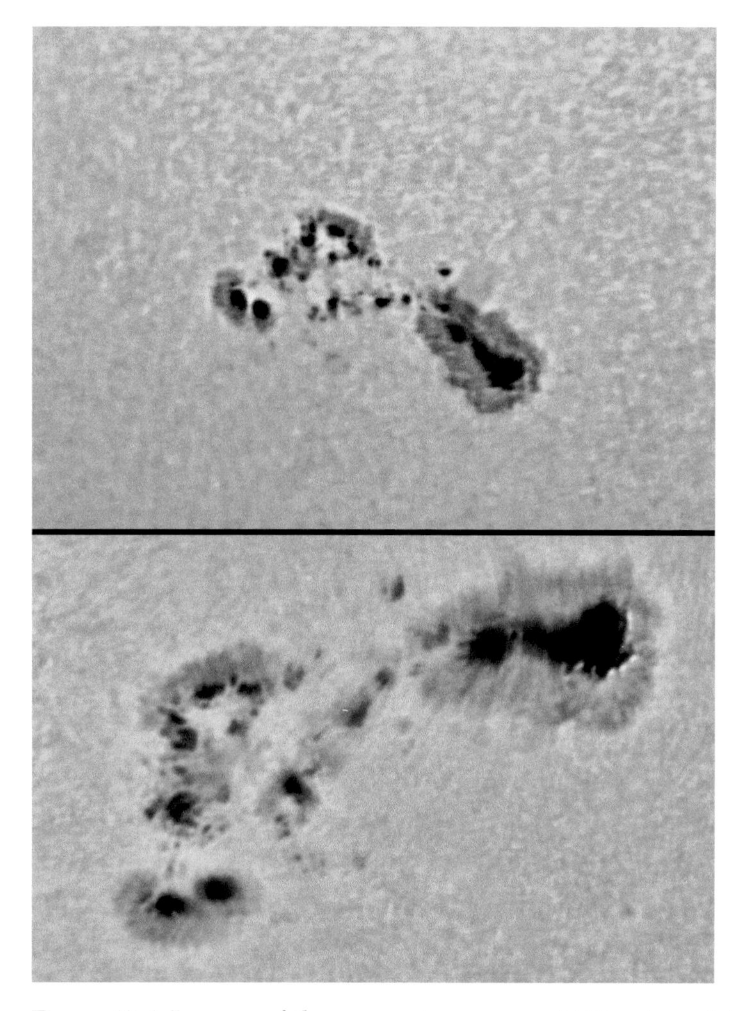

Figure 15.4 Images of the sunspot group no. 2396 using ×2 (upper) and ×4 (lower) Barlow lenses prior to the camera, which was used as a 'webcam' using the *EOS Movie Record* program

local contrast enhancement (using 'Unsharp Mask' with a large radius and small amount) and a little further sharpening using the 'Smart Sharpen' filter in *Adobe Photoshop* to give the final results, which are shown in Figure 15.4.

Having acquired a Baader Herschel wedge two days later, I was able to use the same telescope and camera (in webcam mode) to image the 2396 sunspot group as it neared the Sun's limb. The wedge incorporated a ND3 filter (giving a ×1000 reduction in brightness) along with a Baader Solar Continuum filter having a 10 nm bandwidth at 540 nm. This gives a 'lime green' image (shown in the inset to Figure 15.5) which both increases contrast and eliminates any chromatic aberration. I was very impressed with the amount of detail that was achieved in the resulting image (coloured as above), which is shown in Figure 15.6.

Figure 15.5 Baader Herschel wedge and camera mounted on a 127 mm refractor imaging the Sun. A KAYEM solar finder has replaced the normal 9 × 50 finderscope. The inset shows the 'lime green' colour produced by the Baader Solar Continuum filter

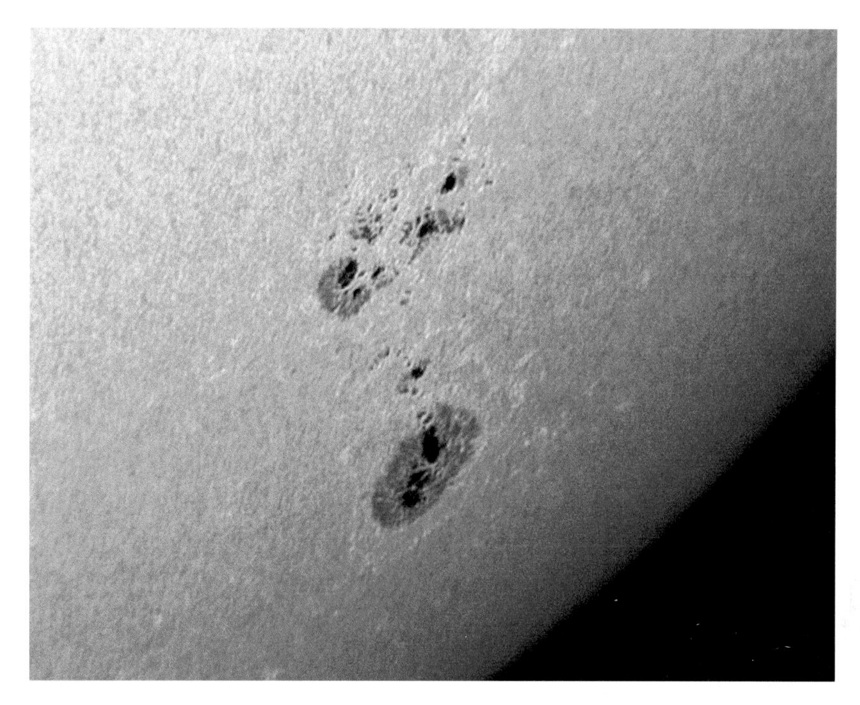

Figure 15.6 Image of sunspot group no. 2396 close to the solar limb taken using a Baader Herschel wedge with ND3 and Solar Continuum filters in the light path

16
Imaging the Sun in the Light of its H-alpha Emission

This does require some significant investment in an H-alpha telescope, but many astronomical societies have one for loan which could allow one to have a first try at imaging in H-alpha. Amateur H-alpha telescopes are made by three companies: Coronado (who are owned by Meade) and Lunt Solar Systems in the USA, and the Isle of Man company, Solarscope. They have to be quite complex in order to isolate the deep red H-alpha line in the solar spectrum which allows great detail to be seen on the solar surface as well as dark faculae lying above the surface and prominences that may be seen at the Sun's limb.

The light passes through a Fabry–Perot etalon composed of two sheets of glass in very close proximity. They produce a comb of very narrow passbands (~0.7 angstroms wide) spaced about 10 angstroms apart across the whole of the visible spectrum. The etalon will reject ~90 per cent of the total sunlight, but this is not sufficient to provide a safe viewing level so that there will also be an energy rejection filter to bring the light level down to a suitable level. This will also block all the infrared light, which is most damaging to our eyes. Following a further filter to reduce the light intensity placed at the front of the diagonal there is a final 'blocking filter' at the exit of the diagonal, which selects only the narrow band centred on the H-alpha line at 6562.81 angstroms, so removing all but the Sun's H-alpha emission.

It is important that the light passing through the etalon is a collimated parallel beam at right angles to the optical axis. The obvious way to achieve this is to place the etalon at the front of the telescope before the converging lens and this is done in many solar telescopes. (As the Sun has a significant angular size, even then the light is not quite a plane wave.) However, etalons are very expensive and an alternative approach that has allowed the cost of solar telescopes to be reduced is to place the etalon within the optical tube where the light cone has become smaller. The light passing through it must still be in the form of a plane wave, so a diverging (concave) lens is placed in front to provide a collimated beam through the etalon, with a second (convex) lens following the etalon used to bring the sunlight to a focus. For example, the Coronado SolarMax II, 60 mm aperture solar scope uses an internal etalon ~40 mm across, reducing its cost by almost 50 per cent. The same approach is used by the Lunt Solar Systems telescope, while the Solarscope range uses a full aperture etalon, which explains why their cost is higher. It is said that with the use of a full aperture etalon it is easier to provide a solar image that has equal detail and contrast across the whole solar disc. I have been able to use a Solaview SV-60 and it does produce a high contrast and well detailed image across the frame. The use of sub-aperture etalons can sometimes give rise to 'sweet spots' across the image making it somewhat harder to produce an image of the whole solar disc.

The effective separation of the etalon's internal surfaces, and hence the precise wavelengths passed, depends on the refractive index of the air between them. This will depend on height and atmospheric pressure, so the effective spacing must be adjusted to compensate. One method – the tilt method – is to vary the angle of the etalon to the incoming light so that it takes a slightly longer path to pass between them (like crossing a road at different angles). This method is used by Solarscope in their Solarview series and is also found in some of the Lunt H-alpha telescopes. A second method, employed in the Coronado H-alpha telescopes, and called 'Rich View', is, I believe, to apply pressure to the etalon (there is a pad at the centre of the etalon) and so alter the separation of the two glass sheets. A third method, used in some of the Lunt telescopes, is to alter the air pressure within the etalon to provide a tuning range of +/– 0.4 angstroms. Thus the etalon always remains perfectly at right angles to the light path, so this highly elegant system is theoretically the best method of etalon tuning. The chosen wavelength may also need to be altered to see at their best the surface detail, faculae lying above the surface or solar prominences as Doppler shifts affect the observed wavelengths.

Many H-alpha scopes can be 'double stacked' to narrow the H-alpha wavelength band and so enable finer detail to be seen on the disc. This narrower bandwidth is obtained by the use of a second, full aperture, etalon placed at the front of the telescope and tuned using the tilt method. By offsetting the two passbands slightly, a slightly dimmer but narrower combined passband is produced. Using two stacked 0.7 angstrom etalons together typically provides a bandwidth of 0.5 angstroms.

Rather than purchasing a dedicated solar telescope, it is possible to buy an H-alpha observing kit to convert a normal refracting telescope into an H-alpha telescope. All the major manufacturers provide etalons with apertures ranging from 40 mm in clear aperture upwards – with larger apertures giving the potential for higher resolution but at a much higher price – along with the required blocking filter, etc. A point to note is that as only a single wavelength is being imaged, the telescope does not even have to be an achromat – it is only important that spherical aberration is well corrected.

Both Lunt Solar Systems and Coronado provide relatively low cost H-alpha telescopes of 35 mm aperture as well as their 50 mm and 60 mm telescopes, seen in Figure 16.1, along with largeraperture versions. At somewhat higher cost, H-alpha telescopes ranging from 50 mm upwards can be obtained from Solarscope. If one looks at the telescope specifications, it is seen that blocking filters of differing sizes are available, being 5, 10 or 15 mm in diameter in the case of the Coronado telescopes with the larger diameter filters costing significantly more. Lunt Solar Systems provide blocking filters of 6 and 12 mm aperture and the Solascope telescopes have one of 19 mm diameter. (Note: To prevent degradation of the filters by moisture, the diagonal should be kept in a plastic bag along with a desiccant pack when not in use.)

Figure 16.1 Coronado SolarMax II 60 and Lunt Pressure Tuned LS60 H-alpha telescopes

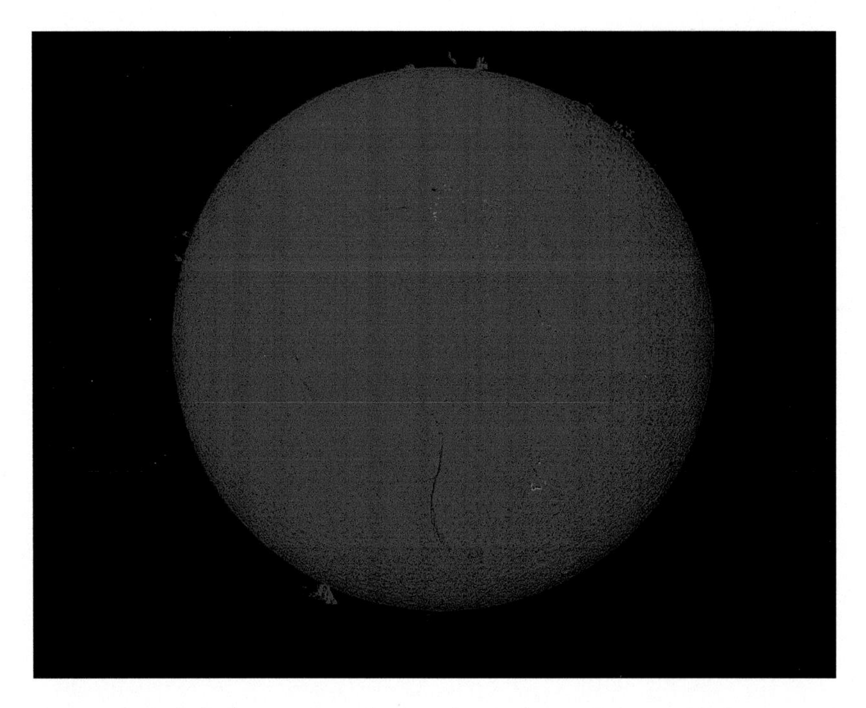

Figure 16.2 Full disc image taken with a Nikon D7000 and Solarscope SV-60 H-alpha telescope

The Lunt Solar Systems and Solarscope telescopes are equipped with TeleVue Sol-Searcher finders, while the Coronado telescopes are equipped with their own Sol Ranger finder. Both project a circular disc (not an image) onto a translucent screen at their rear.

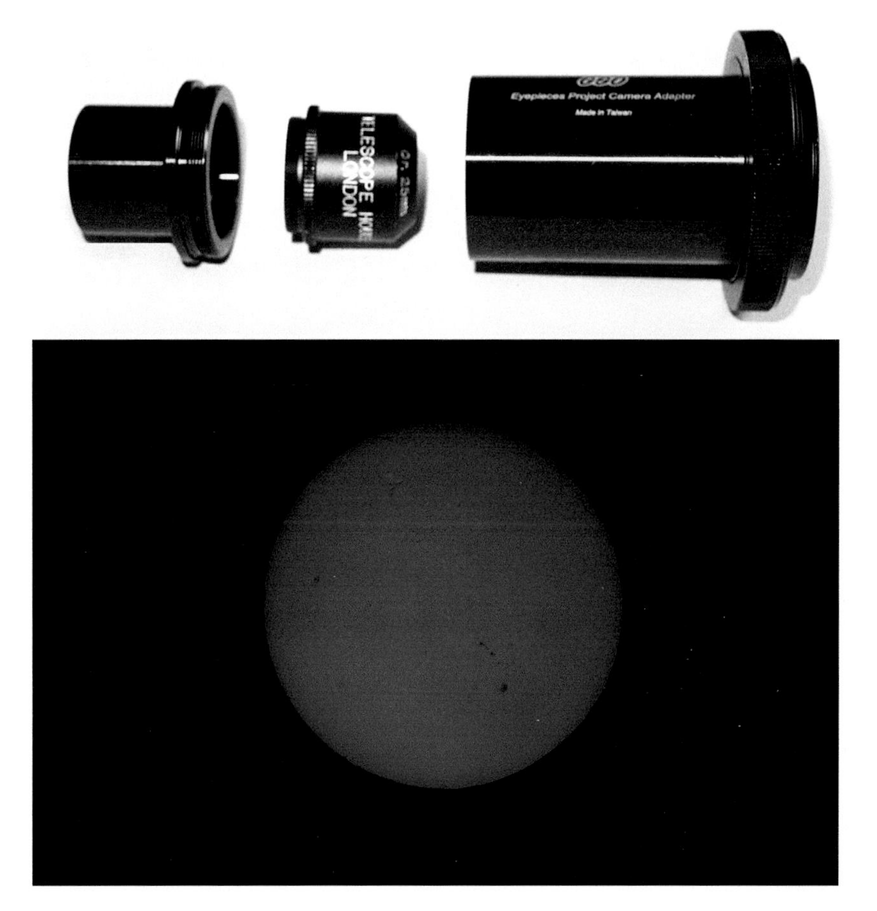

Figure 16.3 Exploded view of a 1.25 inch 'eyepiece projection adapter' employing a 25 mm orthoscopic eyepiece to project the image onto the camera sensor. Below is the resulting image on an APSC sized sensor

Many of these telescopes use a two stage focusing system. The focuser is mounted on an extension barrel which moves in and out of the telescope to obtain an approximate focus. Fine focus is then achieved using a helical focuser in the case of the Coronado, Solarview and Lunt Solar Systems 50 mm telescopes and a Crayford focuser in the case of Lunt's larger telescopes. At greater cost, the latter can be purchased with a very high quality 2 inch Feather Touch focuser. One important point is related to 'focusing' the image. It is necessary to first adjust the focus to give a sharp focus and then 'tune' the etalon to give the clearest image, depending on the atmospheric pressure and the region of the Sun to be imaged.

If using an H-alpha telescope for visual observing, the 5 mm filter is fine for use with most eyepieces, but, if imaging with a DSLR at prime focus (often using a T-mount with a 1.25 inch barrel), the larger diameter filters are suggested if vignetting (darkening towards the image edges) is to be avoided. This is because the camera

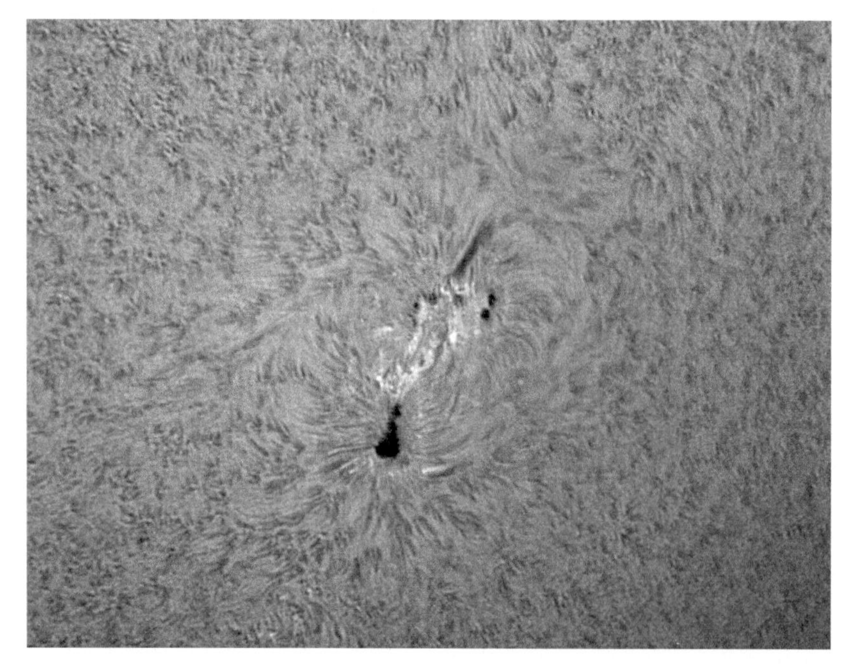

Figure 16.4 A solar active region imaged using a ×2 Barlow and Imaging Source DMK 21AU618.AS monochrome camera

sensor will be at some distance from the filter. Figure 16.2 was taken using a Nikon D7000 DSLR at the prime focus of a Solarscope SV-60 employing a 19 mm diameter blocking filter. The Lunt Solar Systems star diagonal incorporates a very useful feature to aid imaging with a DSLR in that the top of the diagonal incorporates an M42 thread so that given the appropriate bayonet fitment which is part of a T-mount, a DSLR can mounted a touch closer to the blocking filter than otherwise, so reducing any vignetting of the field of view.

Quite often the etalon tuning may need to be adjusted and longer exposures made to show the prominences well, but this will burn out the solar disc. The prominences then need to be added to the full disc image. If the prominences are just visible in the shorter exposure image, it is possible to clone the brighter versions onto their positions – selecting from one image and pasting over the other.

However, *it is possible* to use a DSLR with a 5 mm blocking filter if either a Barlow lens (to image part of the disc) or a 1.25 inch 'eyepiece projection adapter' (to image the whole disc) is used. The latter holds within it a 25 mm orthoscopic or Plössl eyepiece (without its barrel) to 'project' the solar disc onto the camera sensor. An exploded view of an adapter is shown in Figure 16.3 along with an image taken with an H-alpha modified Canon 1100D.

The larger blocking filters will certainly be necessary should a binoviewer be used. The Sun in H-alpha a superb object to observe through the pair of eyepieces.

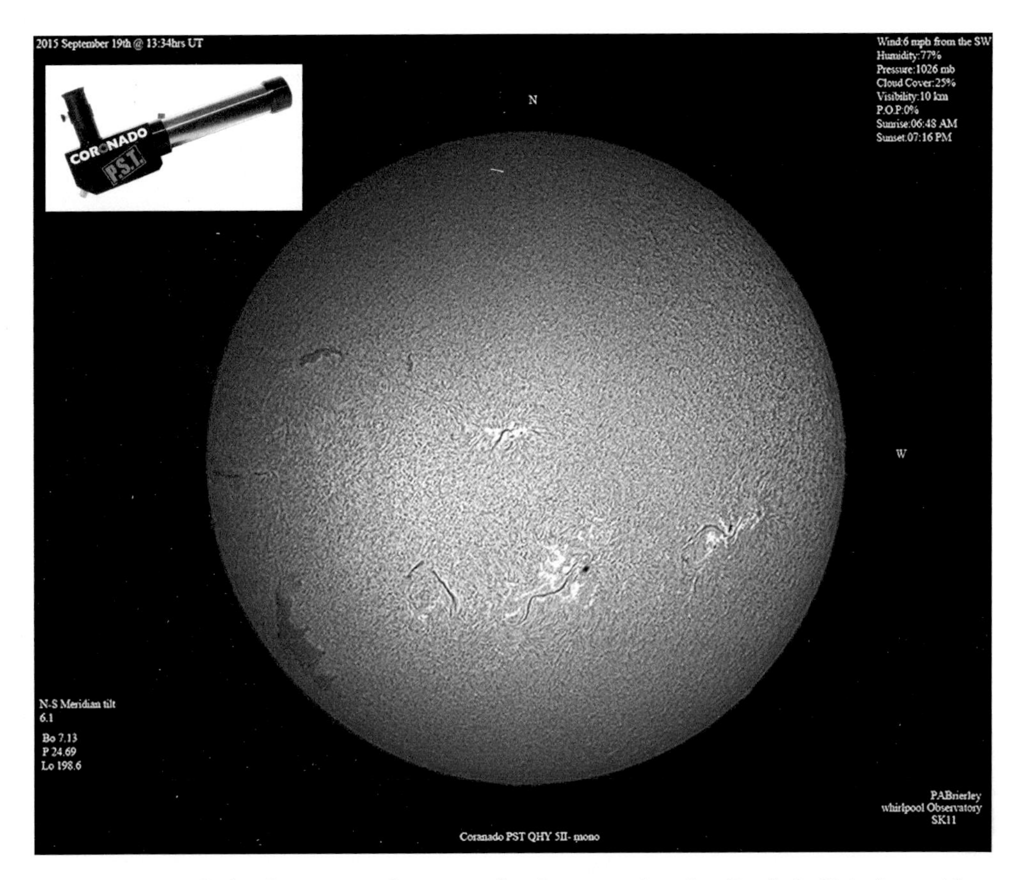

Figure 16.5 Whole-disc monochrome solar image taken by Paul A. Brierley with a Coronado PST (inset) and QHY5II-M webcam

To help overcome the atmospheric seeing, a Canon DSLR can also be used in conjunction with the free program *EOS Movie Record* to produce video sequences that can be processed in *Registax 4*, *Registax 6* or *AutoStakkert! 2*, as has been described in detail in previous chapters. The alternative is to use a 'webcam' such as the Imaging Source DMK 21AU618.AS monochrome camera. In this case, the sensor will be located far closer to the blocking filter, so only a 5 mm diameter blocking filter is required. To image prominent sunspot groups at higher resolution, such as shown in Figure 16.4, a ×2 Barlow lens can be used. A video sequence of 1000 frames was taken using a Coronado SolarMax II 60 and DMK 21AU618. AS camera and loaded into the free software program *AutoStakkert!* 2. Having analysed, aligned and stacked the best images, it produces unsharpened and sharpened images that are loaded into a new folder within that containing the video sequences.

If a monochrome camera has been used, images can be coloured in *Adobe Photoshop* by converting them to RGB mode, duplicating the layer and painting over this

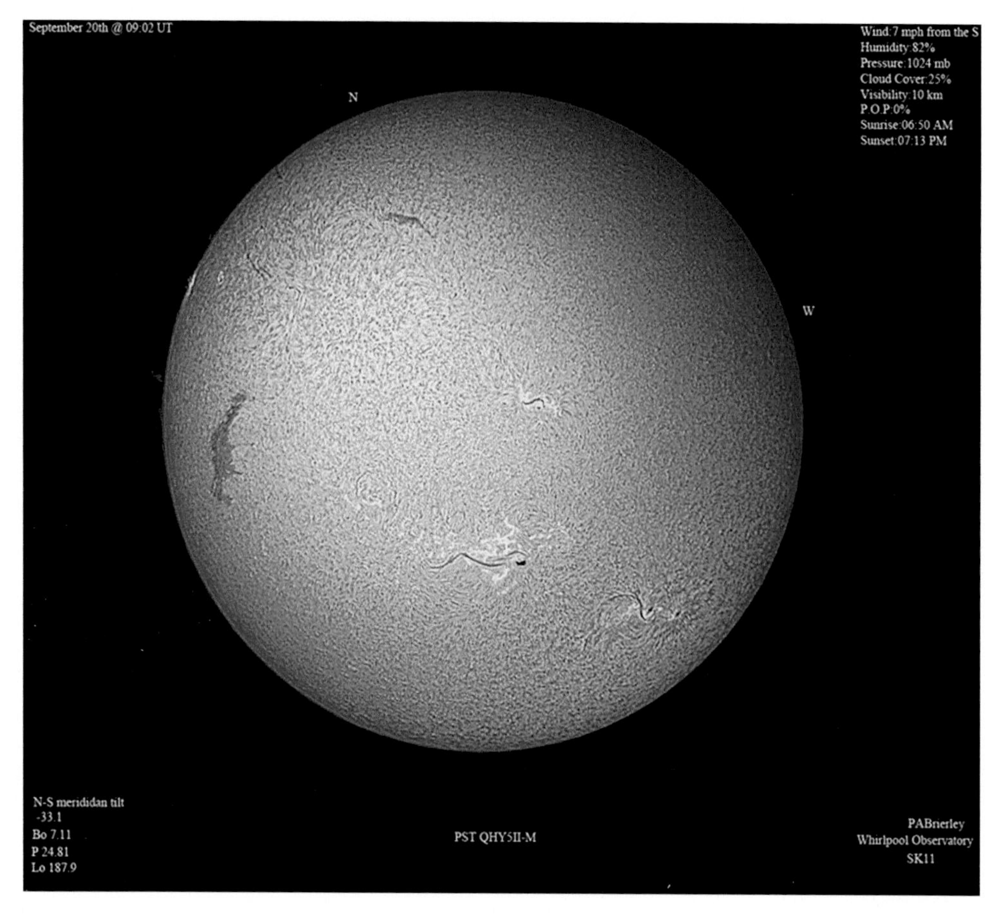

Figure 16.6 Whole-disc coloured solar image taken by Paul A. Brierley with a Coronado PST and QHY5II-M webcam

layer with the paintbrush coloured with R = 244, G = 15 and B = 48 (red) or R = 173, G = 123 and B = 16 (orange) as two possibilities. The two layers are then flattened using the 'Colour' blending mode. The effect is not always good, but sometimes gives a nice result.

As shown in Figures 16.5 and 16.6, both taken by Paul A. Brierley, the Coronado PST (Personal Solar Telescope) with 5 mm blocking filter can give excellent results when using a webcam camera such as, in this case, the monochrome QHY5II-M.

Figure 16.8 is a well annotated solar image taken by Kevin Kilburn using the Lunt Solar Systems 60 mm pressure tuned solar telescope. He has pioneered an interesting technique to show the disc in monochrome – which very often appears better than when coloured – but colouring the chromosphere and prominences. Two videos need to be taken in sequence, one exposed correctly for the disc and

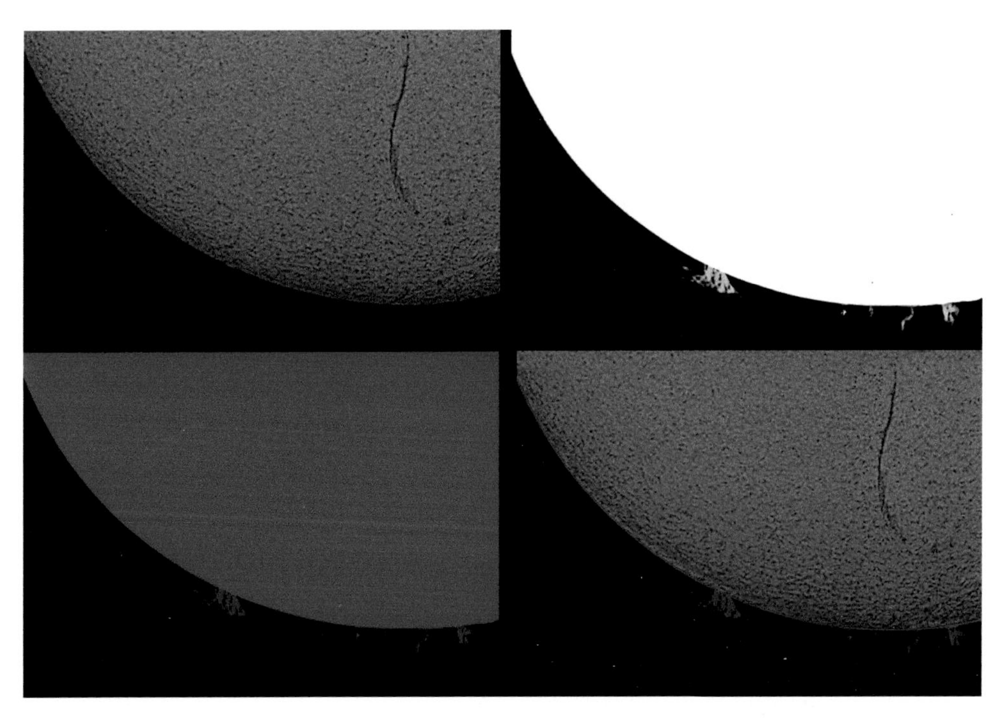

Figure 16.7 Combining solar disk and prominence images as described in the text

a second that will show the prominences well but burn out the disc. Having processed these in *Registax* or *AutoStakkert! 2*, one will have two images, as shown in Figure 16.7, top left and top right. The mode of both images should then be set to RGB. The prominence image is then painted using the colour red (R = 244, G = 15 and B = 48). To achieve this, the image is duplicated and the duplicate layer over painted to give a uniform red colour. The two layers are then flattened using the 'Darken' blending mode to give the image seen in Figure 16.7 (bottom left). To make the chromosphere a little more prominent, the monochrome image of the disc can be reduced slightly in size – perhaps using a percentage of 99.5. The solar disc has then to be cut out with the surrounding black sky made transparent. One way to do this is to duplicate the layer, use the magic wand to select the sky and then hit delete. The disc (in the duplicate layer) will then be surrounded by a grey and white checker board. In the 'Layers' window, the base layer is deleted by dragging it down to the dustbin symbol (bottom right of the Layers window). The cut-out image can be saved if desired as a .psd image without flattening the layers. This image is then copied and pasted over the coloured chromosphere image and, using the 'Move' tool if necessary, aligned with the prominence image. The two layers are then flattened to give the final result as seen Figure 16.7 (bottom right). One nice aspect of solar imaging is that it can generally be done in the warm!

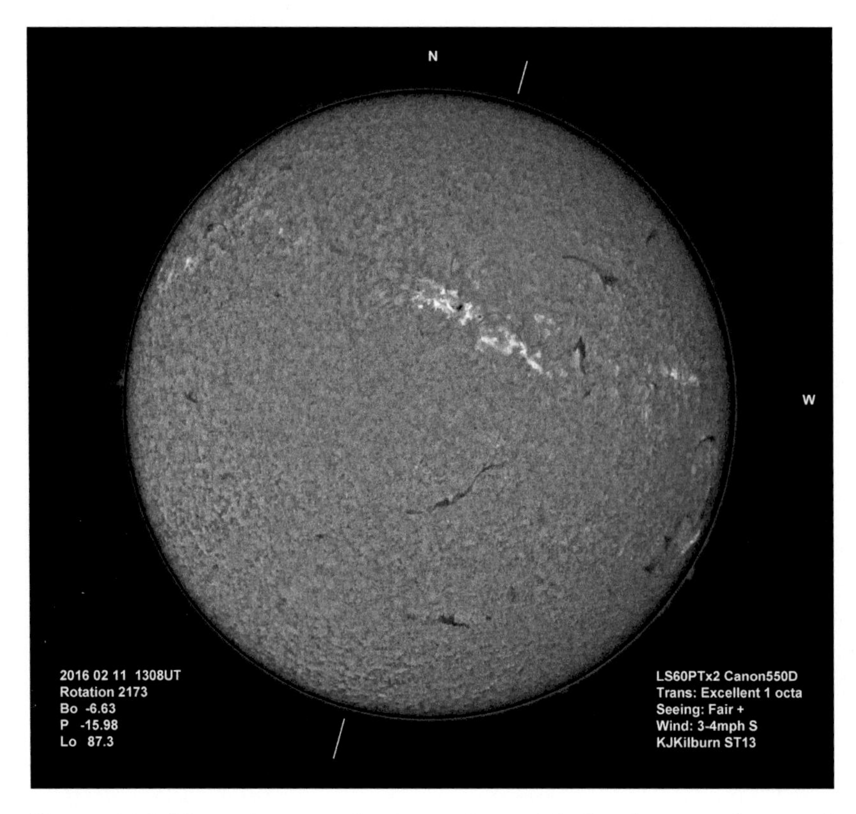

Figure 16.8 Monochrome solar image but with the chromosphere and prominences coloured
(Image Kevin J. Kilburn)

17
Imaging Meteors

In this chapter I will discuss how best to go about capturing meteor trails and then show how to include them within a composite image such as those that can be seen on the NASA APOD (Astronomy Picture of the Day) website (search for APOD Meteor Trails). I assume that a DSLR will be used and suspect that most will have one with an APSC (DX) sized sensor. However, the cost of full frame (FX) DSLR bodies has now fallen to the point where they could, perhaps, be considered. For example, Canon 6D and Nikon D610 bodies can now be obtained for less than £1000 and, having quite a number of Nikon FX lenses, I acquired a D610 largely for use in astrophotography. This has a significant bearing in putting together a system for imaging meteors. The area of an FX sensor is 2.35 times greater in area than a Nikon DX sensor (2.63 times greater than a Canon DX sensor), so, if using a lens that can cover a full frame sensor, one will more than double one's chances of capturing meteors as over twice the sky area will be imaged.

The choice of lens is actually quite a complex question. The sensitivity of a DSLR to meteor trails depends on the area of the lens's aperture. This is 491 square millimetres for a 50 mm, f/2 lens but just 72 square millimetres for a 24 mm, f/2.5 lens and only 6.4 square millimetres for an 8 mm, f/3.5 lens. One might therefore think that the 50 mm lens would be the obvious choice, but there are two other considerations to bear in mind. The shorter focal length lens will see far more of the sky (3.7 and 11 times greater for the two lenses mentioned above) and so give a greater chance of capturing meteor trails. The trail will also move more slowly across the sensor when a short focal length lens is used, allowing the pixels along the trail to collect more light and so increase their effective sensitivity. It turns out that shorter focal length lenses will detect brighter meteors over a wide area of sky, while the 50 mm lens will detect fainter meteors but within a smaller area. There is therefore no obvious 'best' lens to use, though I suspect that a wide angle lens would be a better choice if one was hoping to make a composite image. I have used both a Samyang 8 mm fisheye lens coupled to a Nikon D7000 with APSC sensor to give almost full-sky coverage and also a 24 mm, f2.5, Tamron lens coupled to a Nikon D610 having a full frame sensor.

As meteor trails only last for a few seconds, short exposures using a high ISO will best capture them. The noise performance of the latest DSLRs is quite amazing and ISOs of up to 3200 could be reasonably used. If one uses a very short exposure, say 10 seconds, a vast number of images will need to be taken and the gap between exposures of perhaps 1 second will begin to become significant. I suspect that an exposure of, say, 30 seconds with an ISO of 800 to 3200 would be a good choice. But first try some test images of the sky to ensure that reasonably faint stars can be seen. Given that the large number of images will need to be inspected (a large SD card will be needed), JPEG files are probably the better choice. A dark sky is an absolute must: using a lens at full aperture and high ISO in a light polluted location will give an almost white sky!

Two fully charged batteries will probably be needed or an AC adapter (costing ~£40) used instead. If (as I suspect) one is in a dark location away from mains power, a 12 volt battery and 12 to 240 volt inverter (costing ~£25) would also be needed if an AC adapter were used. An intervalometer to trigger each exposure is also a real necessity. Some cameras include them within the system, but I find them fiddly to use, particularly in the dark, and much prefer to use an external intervalometer. These can be bought for well under £20, but it is important that one with the correct connector for the DSLR is chosen. The camera is set to manual mode and the exposure time and lens aperture set. If the exposure were 30 seconds long, the intervalometer would be set to initiate an exposure every 31 seconds. My Viltrox intervalometer allows for a total of 399 sequential exposures, so allowing for a continuous imaging period of over three hours. The camera should be mounted on a sturdy tripod and aimed around 30–40 degrees away from the radiant (near the radiant, the trails are very short as the meteors are coming towards you) and at an elevation of ~50 degrees as this captures the height in the atmospheres where most trails are to be seen. To prevent star trails, a tracking mount, in this case a Baader Nano Tracker, would be a useful asset as seen in Figure 17.1 but this is not too necessary if a wide angle lens is used. (However, it would be a great help if a composite image is to be made, as described below.)

A lens will tend to dew up after about an hour and so must be kept warm. As described in Chapter 1, a dew strap could be used, but one can simply place a short sock around the lens, with a hole cut in the toe, to hold a pair of hand warmers. A neater, but more expensive, alternative is to purchase a LensMuff from Kevin Adams in the USA. (Search for 'LensMuff'.) These can hold three hand warmers and have a Velcro strap to hold them around the lens.

The use of a short exposure and high ISO will naturally give a rather noisy image and so a single image of a meteor trail will not be too good. This is not too important if one is going to try to produce a composite image of several meteors captured during a meteor shower. Figures 17.2 and 17.3 show an image of a meteor taken with the Nikon D610 at an ISO of 1600 and Tamron 24 mm, f/2.5 lens at full aperture mounted on the Baader Nano Tracker.

Exactly the same techniques to enhance this image were used as those used to produce the Perseus and Cassiopeia images as discussed in Chapter 2 and the removal of light pollution that was described in Chapter 12. (The image was

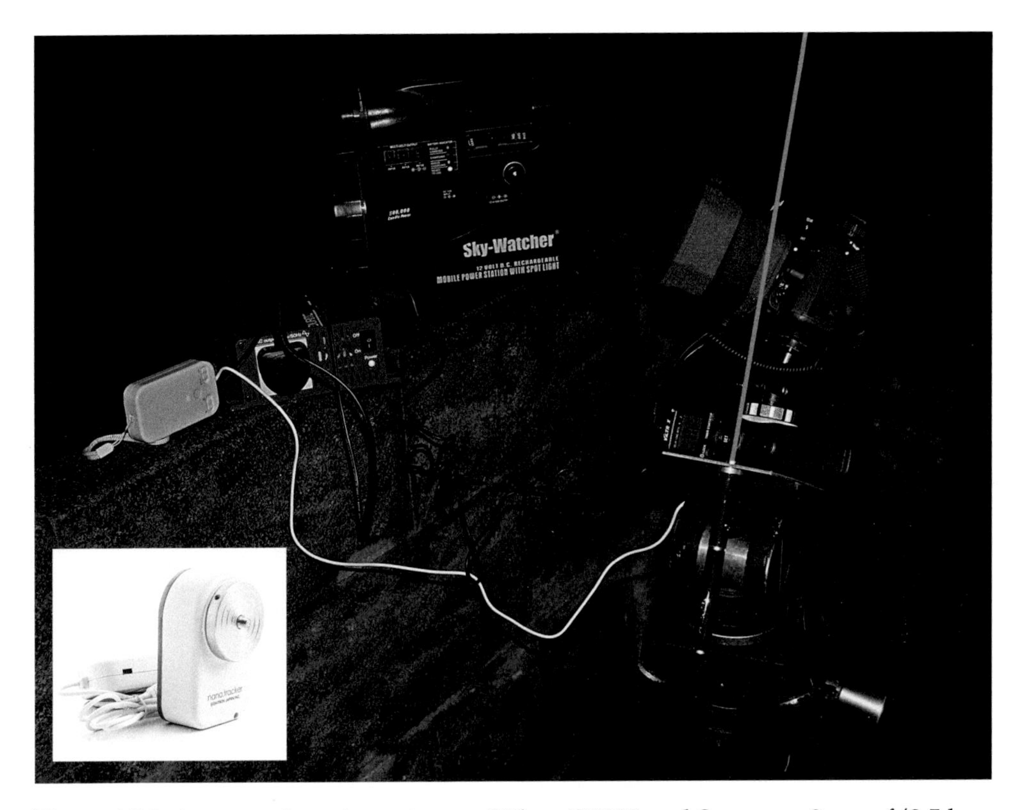

Figure 17.1 A meteor imaging set-up: a Nikon D7000 and Samyang 8 mm, f/3.5 lens mounted on a Baader Nano Tracker. To prevent dew forming on the lens, a LensMuff surrounds it. A SkyWatcher Power Tank provides 12 volts to an inverter to give the 230 volt supply for an externally powered Nikon 'dummy battery'. A laser pointer is mounted on a metal plate to enable the Nano Tracker to be polar aligned and a Viltrox intervalometer is used to initiate the 30-second exposures

stretched using a few applications of the 'Levels' command with the middle slider set to 1.2. This showed some light pollution towards the lower part of the image and this was removed by duplicating the base layer applying the 'Dust and Scratches' filter with a pixel size of 10 pixels and cloning out any remaining bright stars (and in this case, the Orion nebula) to produce an image of the light pollution. Flattening the two layers using the 'Difference' blending mode then removes the light pollution. The image was further stretched, the brighter stars selected using the colour range tool and the selection expanded by 9 pixels. The Gaussian Blur tool was then used with a radius of 4 pixels to expand these stars, but as this make them fainter, their brightness was brought back using the 'Curves' tool, with the fainter parts of the image 'pinned' but the brighter parts lifted.)

The stars of the main constellations can be seen along with Hyades and Pleiades open clusters, but the image is very 'blotchy' – a common problem with DSLR

Figure 17.2 A Geminid meteor imaged with a short exposure and showing some light pollution towards the bottom of the image

sensors as the sensitivity of the different colour pixels tends to vary across the sensor. Pleasingly, both the red giants Betelgeuse and Aldebaran show their orange colour, but this, in itself, is not a good image. I can only assume that the far better images such as that imaged by my colleague Robin Skagell (shown in Figure 17.4) and those that can be seen on the Internet have used longer exposures so that the images are less noisy. As will be seen below, if a composite image is to be made, the quality of the star field against which the meteor trails will be seen does not matter as they will be 'added into' a high quality star background. Figure 17.9, below, shows the same meteor trail seen against a far higher quality star field produced at the same time.

Imaging a Meteor Shower

The beautiful images of meteor showers on the 'astronomy picture of the day (APOD) NASA website show many meteor trails against a beautiful star background, while also having a smooth sky background and attractive foreground. These very well crafted composite images are not too difficult to achieve, as I hope this part of the chapter will demonstrate. The procedure does require a reasonable working

Figure 17.3 The enhanced image of the Geminid meteor trail – noisy owing to the very short exposure

knowledge of *Adobe Photoshop,* but many of the processes outlined below have been covered in more detail in previous chapters.

The first requirement is to produce an image of the sky where one expects to observe the meteors. As an example, let's take the Geminid shower in mid-December. Gemini lies in an interesting region of the heavens close to the Auriga, Orion and Taurus constellations, so an area of the sky including them would be a good region to choose. Using the *Astronomy Tools* field of view calculator, I found that a 24 mm lens when used with a full frame DSLR sensor would nicely encompass this region and was then able to purchase a second hand Tamron lens of this focal length.

So, for this demonstration, this region was imaged with a Tamron 24mm, f/2.5 lens stopped down to f/5.6 (to give better image quality than when at full aperture) coupled to a full frame Nikon D610 camera at an ISO of 800. A set of twenty-four 25-second exposures were made giving 10 minutes total exposure. A tracking mount is a great help and a Nano Tracker was used for these exposures as seen in Figure 17.1, but otherwise exposures of 10 seconds will minimise star trails. As many images are to be combined, JPEG capture would be adequate, but raw files could, of course, be used. The images were stacked and aligned using *Deep Sky Stacker* and then stretched with two applications of the Levels command with the centre slider set to 1.2. The same

Figure 17.4 A superb image of a Geminid meteor taken by Robin Skagell

Figure 17.5 The 'Skyscape' image which was the result of stacking twenty-four 25-second exposures

techniques used in Chapters 2 and 12 and summarised above were used to stretch the image, remove the small amount of light pollution and enhance the brighter stars. I called the resulting image, which shows little noise and whose main constellation stars have been enhanced, 'Skyscape'.

Totally black skies are not too attractive and the Skyscape image could have some colour added. In this case, the image was saved as 'blue sky' (to ensure that the skyscape image would not be lost) and over painted with a blue colour (R = 56, G = 82, B = 75). This gives a blue image of the same pixel dimensions. As skies tend to get brighter towards the horizon, the gradient tool was then used to darken this towards the top. This is shown in Figure 17.6.

The result can then be added into the skyscape using the 'Normal' blending mode with a suitable opacity (56 per cent in this case). Some experimentation is required. The final touch might then be to add an interesting foreground to the base of the image. The canvas size (in black) of the skyscape was extended beneath the image and a suitable foreground image copied and pasted over it. This could even have been taken in daylight and suitably darkened first. The appropriate blending mode would depend somewhat on the image to be added, but the 'Lighten' mode is likely to work well. One might well consider that this is 'cheating', but I promise you that

Figure 17.6 A graduated 'blue sky' having the same dimensions as the skyscape image

this is how these composite images are made. This is quite a pleasant image even without the meteors added to it.

The requirements for best capturing meteor trails during a shower were outlined above. Ideally the same camera and lens employed to make the skyscape should be used, preferably on a tracking mount and having the same orientation on the sky as this will make adding the trails into the skyscape easier. In this case the exposures to make the skyscape image were taken prior to opening the lens's aperture and increasing the ISO to 1600 in order to, hopefully, capture some of the Geminid shower. Exposures of 30 seconds were used with each exposure triggered every 31 seconds. As seen above in Figure 17.3, the star images will not be good but this is not important; neither is any star trailing that would result should you not have a tracking mount as the effective exposure of the meteor trail is only a few seconds.

Having looked through them all (a good reason for using JPEG files), one hopes that a few will show a meteor trail. The position of the trail with respect to the stars in the field should be noted and then the region around it cropped. (In this first example, the Hyades Cluster was adjacent to the meteor trail.) Using the 'Polygonal Lasso' tool, the trail is selected as shown in Figure 17.8, top. The selection is inverted and the remainder of the field painted black as shown in Figure 17.8, bottom.

Figure 17.7 The final skyscape image with a foreground added

This image of the meteor trail should then be copied and pasted over the Skyscape image with the blending mode set to 'Lighten'. (This compares each pixel of the two layers and puts the brighter of the two into the flattened image – perfect for this process.) The trail will now be seen within the background stars. By pressing 'Ctl-T', the trail can be moved using the cursor keys to its correct position in the field and, if necessary, rotated so that it points away from the radiant – in this case just above the star Pollux.

Exactly the same process is employed for all the trails that have been captured to give a composite image of meteor trails against the star background shown in Figure 17.10. The meteor trails could have been captured during more than one shower, and in this case the majority were captured near the equator during a wonderful display in December 2015 which included several 'near' fireballs.

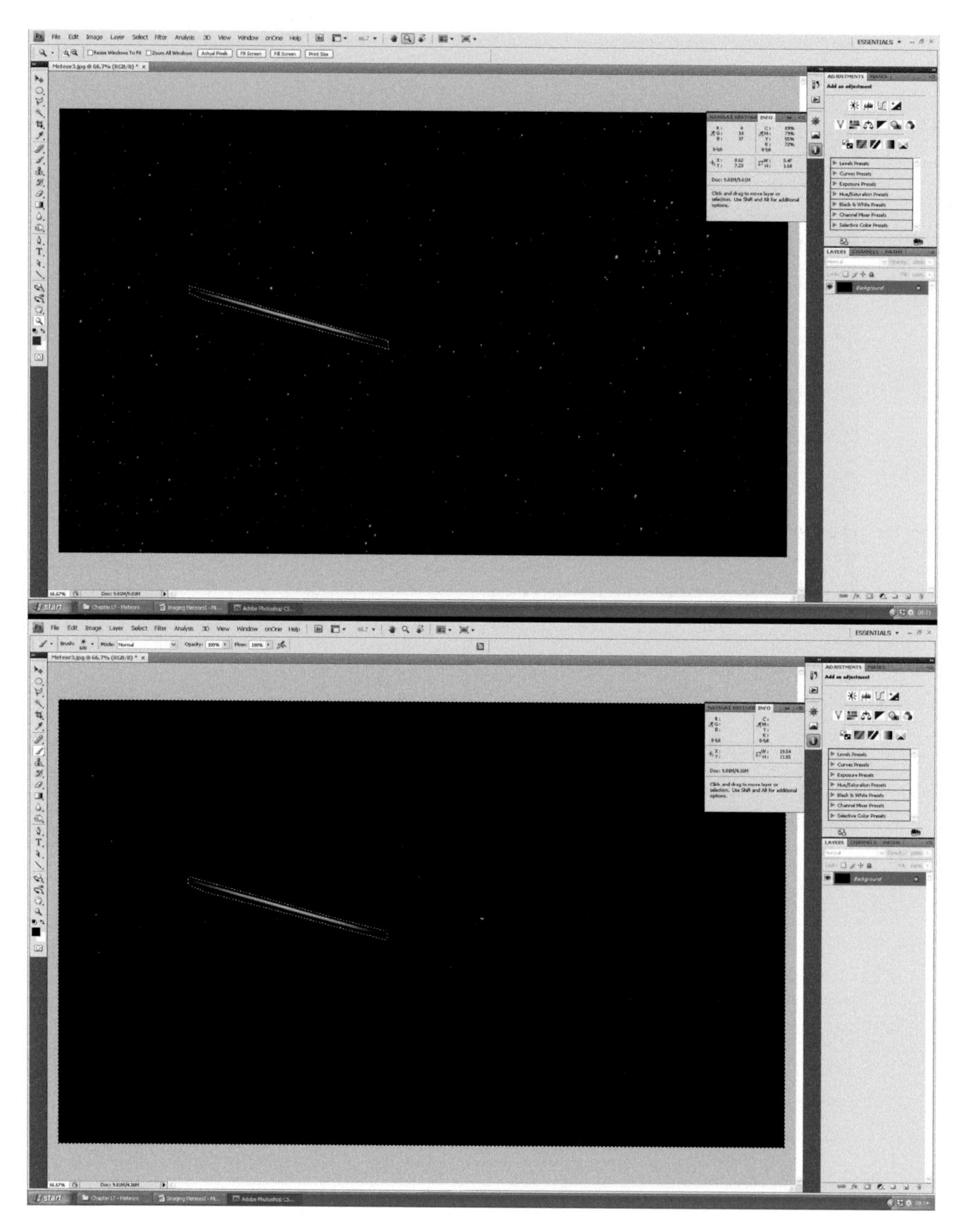

Figure 17.8 Selecting the meteor trail from an image (top) and the other parts of the image painted black (bottom)

Figure 17.9 The meteor trail shown in Figure 17.3 added into the skyscape image

Figure 17.10 A final composite image showing some Geminid meteors

18
Imaging Comets

Comets can be some of the most beautiful objects to observe and image in the heavens, but their images need a little care in processing owing to the fact that they are often moving quite quickly relative to the stars. Very bright comets, those easily seen with the unaided eye and which usually subtend a reasonable arc across the sky, will be imaged best with a DSLR camera and 50 to 200 mm lens. Given their relatively long focal length, a tracking mount (such as described in Chapter 3) is very useful, but otherwise a number of very short exposures can be stacked in *Deep Sky Stacker*. Comets that reach magnitude +5, and so could just be seen under dark skies, are termed bright but may well not be easily visible under typical skies. These would probably be imaged with a short focal length refractor on an equatorial mount. Some help is then required to find comets. Astronomy magazines will usually give a star chart showing how to locate them. An even better alternative is to use a planetarium program such as *Sky Safari* (search for 'Southern Skies Sky Safari') which automatically downloads the orbital elements of new comets and displays their precise location in the sky on any given date. The 'Plus' version will also control a wide range of telescope mounts, and so, having aligned and synchronised the telescope on a nearby bright star, clicking on the comet will centre it in the field of view.

Having acquired the comet, exposures similar to those used for imaging star fields would be used. If using a DSLR, I would tend to use 30-second or 1-minute exposures with an ISO of 800. It is important to make some test exposures to ensure that the comet's coma is not overexposed. Owing to the comet's motion against the sky, tri-colour images taken with a monochrome CCD camera are somewhat difficult to process and a colour CCD camera would be a better choice.

The individual images (frames) taken of the comet must then be aligned and stacked and herein lies the problem. If one aligns on the stars, the comet will become blurred owing to its motion against the stars, and if one aligns on the comet, the stars will trail. The solution is to align and stack on each individually and replace the blurred comet in the stars image with that from the stacked comet image. Happily, the program *Deep Sky Stacker* will do this for you automatically. The procedure is, as one might expect, a little more complex than a normal

Figure 18.1 *Deep Sky Stacker* had aligned on the stars within the 10 images giving a somewhat extended comet image

stacking process. As a demonstration as to how this is done, let's suppose that the comet is quite bright and only 10 exposures are needed to show it well. These are loaded into *Deep Sky Stacker* as normal, along with dark and flat frames if one feels that they are needed. (Dark frames taken at the same exposure and ISO will help remove hot pixels and amplifier glow, and flat frames will correct for any vignetting towards the corners of the field.)

I have found that it is best to first stack the frames as normal, remembering that you may need to open the 'Advanced' tab within the 'Register checked pictures' box and adjust the star detection threshold. This provides *Deep Sky Stacker* with alignment data for each of the frames and provides a sharp star image which when stretched and enhanced provides an image of the comet, as seen in Figure 18.1, whose coma can be seen to be extended owing to its motion against the stars.

Deep Sky Stacker cannot align on the diffuse comet images, so this has to be done manually for each frame. The first light frame in the list is clicked upon, when, to the right of the screen, a set of symbols appears, one of which is a comet (arrowed in red) as shown in Figure 18.2.

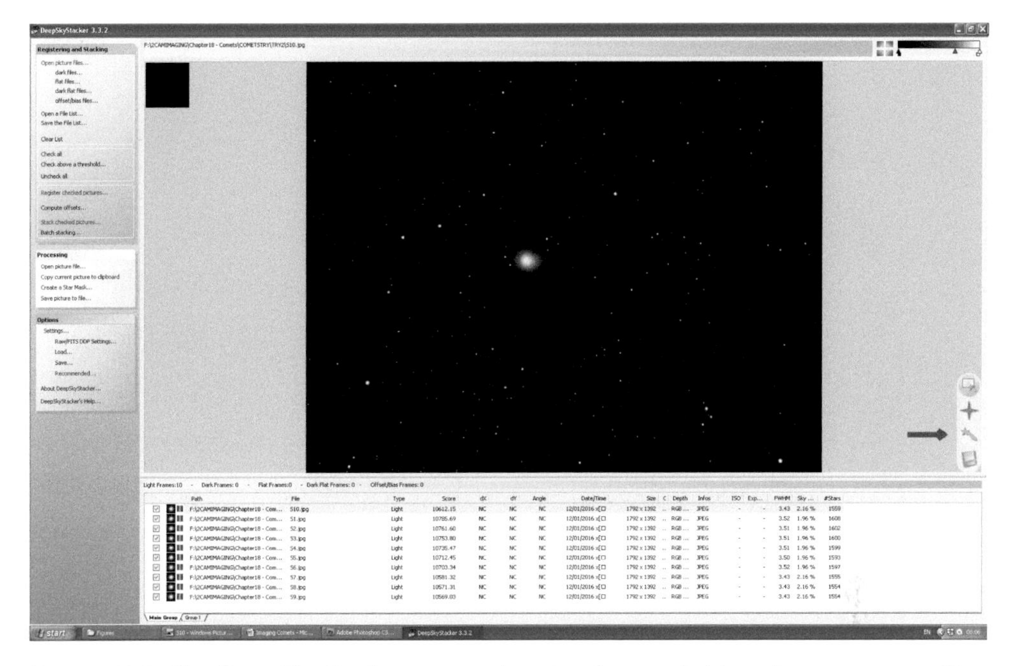

Figure 18.2 The *Deep Sky Stacker* screen showing how to initiate the process to align on the comet's coma

The 'comet symbol' is clicked upon, the cursor is centred on the comet's coma and, *with the shift key pressed*, it is clicked upon. A mauve circle will appear within the coma and *Deep Sky Stacker* then knows its position within the frame, as seen in Figure 18.3. This is repeated for each of the light frames.

The 'Stack Check Pictures' box is then opened and at its top left it says '**Stacking Mode:** Standard'. The word 'Standard' is clicked upon, whereon a box opens up with a 'Comet' tab at the top, second from the left. When this is clicked, a new box opens as seen in the Figure 18.4 which gives three options.

It is useful to then implement the comet stacking mode, which will provide a sharp comet but a pretty odd star field (owing to the way *Deep Sky Stacker* combines the images) as shown in the top image of Figure 18.5. Finally, the 'Stars + Comet stacking' mode should be used which combines the two to give an image of the comet against a sharp star field, as seen in the bottom image of Figure 18.5. The *Deep Sky Stacker* tutorial states that '*A first stack is created to extract the comet from the background. Then a second stack is created having subtracted the comet from each frame before stacking. The final image is then obtained by inserting the comet back in the image*'.

However, on the trial set of frames used for this processing exercise, the coma in the 'comet with star trails' image (Figure 18.5, top) appeared sharper than that in the combined image produced by *Deep Sky Stacker* shown in Figure 18.5, bottom. I thus combined the two by erasing the comet's coma from the 'stars' image (Figure 18.1)

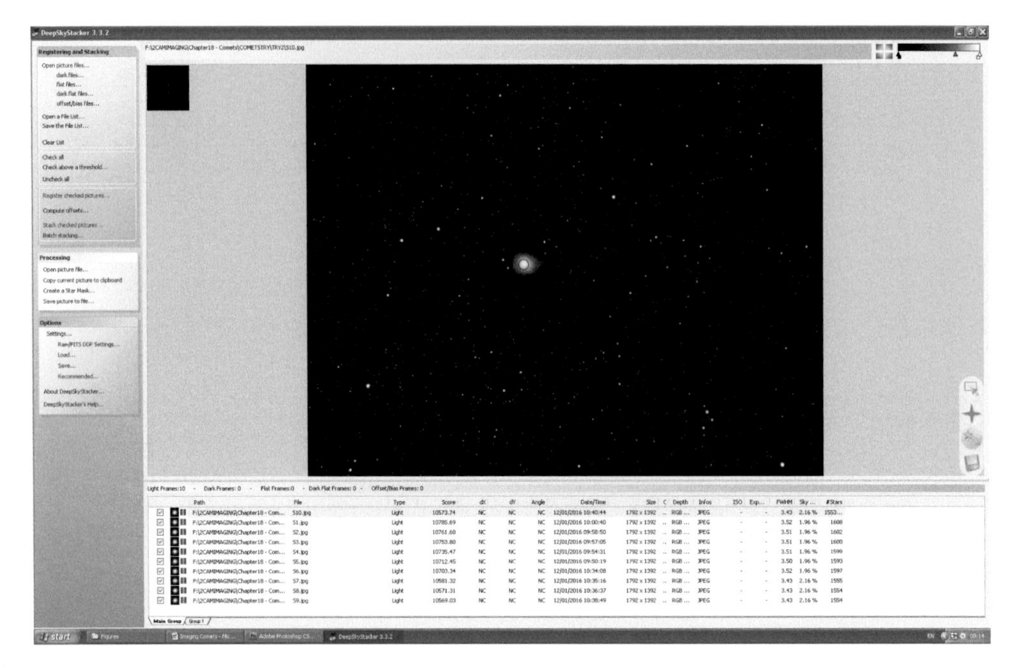

Figure 18.3 Giving *Deep Sky Stacker* the coordinates of the comet's coma in the first of 10 frames

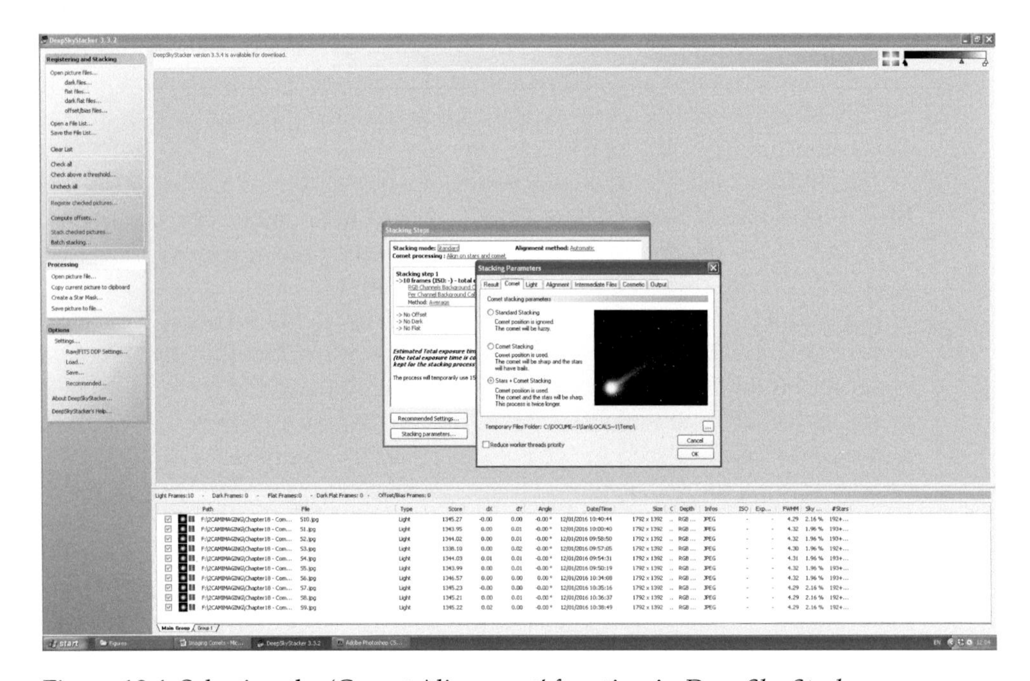

Figure 18.4 Selecting the 'Comet Alignment' function in *Deep Sky Stacker*

Figure 18.5 The results of using *Deep Sky Stacker* to align on the comet's coma (top) and the combination of a sharp star field and the comet (bottom)

Figure 18.6 A composite image of the two images stacked by *Deep Sky Stacker* having first aligned on the stars and then the comet's coma

and cloned the comet's coma from the 'comet' image (Figure 18.5, top) into its location to give the final result as shown in Figure 18.6.

Figure 18.7 is a beautiful image of comet C2013 US10 (Catalina) taken by one of the world's top astrophotographers, Damian Peach, on the night of 6 January 2016 when the comet passed by the globular cluster M3. His website is: www.damianpeach.com.

Figure 18.7 An image of comet Catalina (C2013 US10) imaged by Damian Peach as it passed by the globular cluster M3

19
Using a Cooled 'One Shot Colour' Camera

Although DSLRs can, and do, make great astrophotographs, there may well come a time when one might consider a cooled CCD camera. Their great advantage lies in the word 'cooled' as, by reducing the sensor temperature to typically –20° Celsius (and so perhaps 40 degrees below that of a DSLR sensor) the dark current made up of thermal electrons is reduced by a factor of more than 100 and its noise level by a factor of ~10. Images will thus contain far less noise and longer exposures are made possible as the greatly reduced dark current will not fill the pixel wells so quickly. A further advantage over an unmodified DSLR is that the sensitivity to the deep red H-alpha emission will be significantly greater.

CCD cameras can use either monochrome or colour sensors. The latter has a Bayer matrix of red, green and blue filters above each set of four pixels so that colour images are immediately captured. In contrast, monochrome cameras need to be used with a set of RGB colour filters, so a sequence of image captures is required. The use of a monochrome camera and filters will be discussed in detail in Chapter 20. Many web discussions relate the relative merits of the two types and advanced imagers will tend to use monochrome cameras, but there is no doubt that a colour camera is simpler to use and, as a set of filters is not required, will cost significantly less. One great advantage is that a colour image can be acquired in a single imaging period and so even if clouds roll in after an hour or so, one will still have data to process into a colour image.

Having perhaps been used to capturing images with a DSLR having an APSCsized sensor (as will have many beginners to astrophotography), one is unlikely to want a camera with a significantly smaller sensor. There are then two possible approaches to obtaining a cooled colour camera having an APSC sensor. The first is to make or buy a modified DSLR camera. The cheapest approach would be to buy the Canon EOS D700, 18 megapixel, camera and a cooling modification kit from JWT Astronomy in the Netherlands. This is probably not for the faint hearted! Also based on the D700 is the PrimaLuce Lab D700a (Figure 19.1) which is modified to detect far more of the H-alpha emission, as discussed in Chapter 7, and has an integral cooling unit to cool the camera sensor to up to 30° Celsius below the ambient temperature. It is supplied

Figure 19.1 The PrimaLuce Lab 700Da cooled sensor DSLR camera

with a dummy battery powered from the unit. As well as allowing a defined sensor temperature to be set, the control unit also enables one to program the number and length of exposures that will be taken with the camera so that an external remote control is not required. Very usefully, it includes an anti-dewing system to prevent condensation – or frost – forming on the sensor. This is a very cost-effective way of obtaining a large sensor, high resolution, 'One Shot Colour' (OSC) camera.

The alternative is to purchase a custom made astronomical camera incorporating a colour APSC sensor. The lowest cost of these OSC cameras currently on the market is the QHY QHY8L costing around £1000 as shown in Figure 19.2. The camera uses a 6 megapixel Sony sensor employing 7.8 micron square pixels and, most importantly, incorporates 'set point cooling' to cool the sensor to a specified temperature such as −20° Celsius. This makes the taking of dark frames at the same sensor temperature as the light frames far easier. The camera uses 16-bit digitisation and provides for 2×2 and 4×4 (monochrome only) binning modes. The 4×4 mode, which averages over 16 pixels, is significantly more sensitive and is used for rapid initial alignment on the selected object. The sensor is mounted in a slim body just 63 mm across, so making it perfect for use with a 'HyperStar' imaging system when the camera is mounted in the centre of the corrector plate of a Schmidt–Cassegrain telescope. To achieve this, the multiple power supplies required for the sensor electronics and Peltier cooling system are contained in a small separate unit connected to the camera with a multi-core cable. This is powered by a supplied mains adapter or 12 volt battery taking up

Figure 19.2 The QHY QHY8L cooled, one shot colour camera

to 4 amps current. As seen in Figure 19.2, the USB2 type B and power supply sockets are at the rear of the camera. As is often found with the USB2 type B connector, its placement in the socket of the QHY8L is not very secure. To overcome this problem, as also seen in Figure 19.2, a bracket (made using 3D printing) is supplied to mount on the camera in order to hold the connectors tightly in place. At somewhat greater cost, QHY manufacture a 10 megapixel version to give increased resolution. The ability to use the monochrome binning modes for initial alignment on faint objects makes these cooled CCD colour cameras – and those made by other manufactures such as the SBIG STF-8300C (which is the colour version of the STF-8300M monochrome camera that will be discussed in Chapter 20) – somewhat more practical than modified DSLR cameras. Many employ the rather smaller (17.96 × 13.52 mm), 8 megapixel, Kodak KAH-8300 CCD array, which has superb low noise characteristics. The Moravian Instruments G2-8300 Colour CCD Camera has a very similar specification as does the Atik camera's 383L+.

The *EZcap* capture software provided with the QHY sensors allows a sequential set of exposures to be taken which must be un-binned (1×1) to allow for colour capture. Though these can be exported in the JPEG format, only 8 bits per colour will then be saved; in order to retain the full 16 bits of data per pixel, images must be saved using the FITS format. This is a raw file format which saves a 16-bit word for each pixel but contains no colour information. The software processing the data must thus be told the format of the colour filter matrix in order to 'de-Bayer' the raw data to produce a colour image. When the set of FITS images to be aligned and stacked is loaded into *Deep Sky Stacker*, the 'FITS Files' section within the 'Raw/FITS DPP Settings' window must be opened and the appropriate Bayer matrix employed by the sensor selected ('GBRG' in the case of the QHY8L). *Deep Sky Stacker* then appropriately de-Bayers the monochrome images to give each colour a 16-bit depth in the output TIFF file.

Imaging the Perseus Double Cluster

I imaged the Double Cluster following a several hour session imaging the Andromeda Galaxy, M31, (described below) for which my chosen telescope was a 72 mm William Optics Megrez refractor having a focal length of 432 mm which was used with a Teleskop Service 2 inch field flattener and an Astronomik CLS light pollution filter employed as the cluster was being imaged from a light polluted location. It was mounted in parallel with an 80 mm, f/5 refractor and QHY6 guide camera to apply guide corrections (using *PHD* guiding software) to an Astro-Physics Mach1 GTO mount.

These star clusters are relatively bright and so, if using a cooled colour camera, a total exposure time of ~30 minutes is sufficient. Using the QHY8L, the *EZcap* software was set to take six 6-minute exposures, giving a total exposure of 36 minutes. It is always useful to view the individual images to check for any faults such as guiding errors and hence trailed stars. This can be done with the free program *IRIS*, which will produce a monochrome image when a FITS file is loaded. The set of FITS files were imported (with the appropriate GBRG deBayer matrix selected) into *Deep Sky Stacker* for alignment and stacking along with a set of dark frames. These, taken with the same sensor temperature and exposure time, allow hot pixels and some amplifier glow in the extreme top left of the image to be removed. The Double Cluster can also be easily captured with a DSLR but then I would take short exposures (~30 seconds), to account for the higher dark current, and increase the total exposure time to 60 minutes.

The aligned and stacked image produced by *Deep Sky Stacker*, as seen in Figure 19.3, was exported as a 16-bit TIFF file into *Adobe Photoshop* and stretched using four applications of the 'Curves' function, as shown in Figure 19.4. This lifts the brightness of the fainter stars but not that of the brighter ones.

To reduce the background noise in the image, the 'Levels' command was then used to bring up the 'black point', as seen in Figure 19.5. Perhaps owing to the greater sensitivity of the sensor to green, bright stars showed a green cast.

This was removed by an application of the filter 'Hasta La Vista, Green!', as seen in Figure 19.6. This filter is provided within a free, downloadable, filter set for *Adobe Photoshop* found at www.deepskycolors.com.

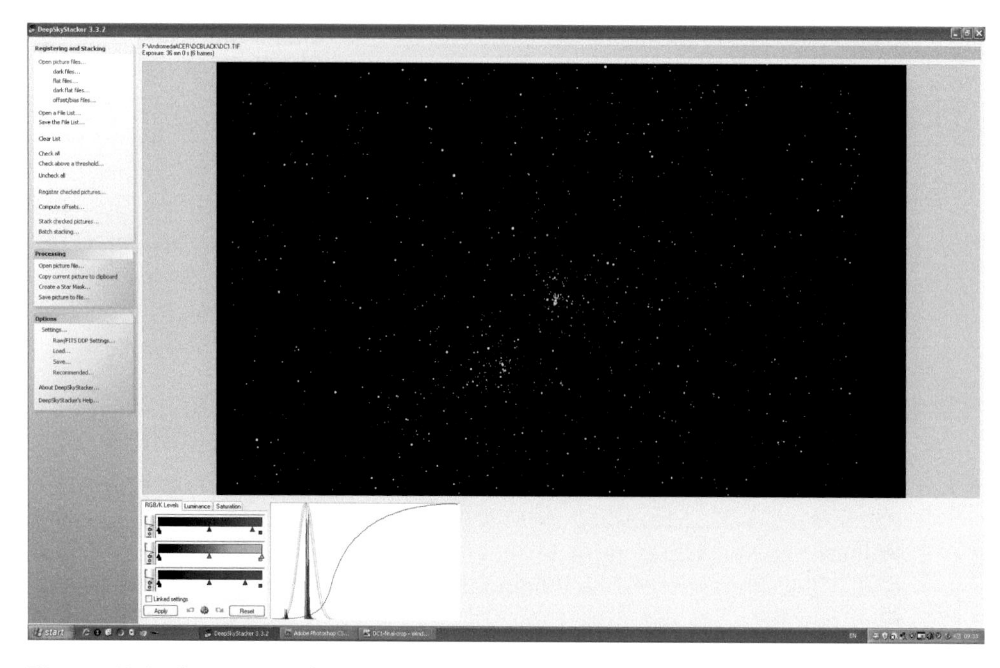

Figure 19.3 The output from *Deep Sky Stacker,* having stacked six 6-minute guided exposures

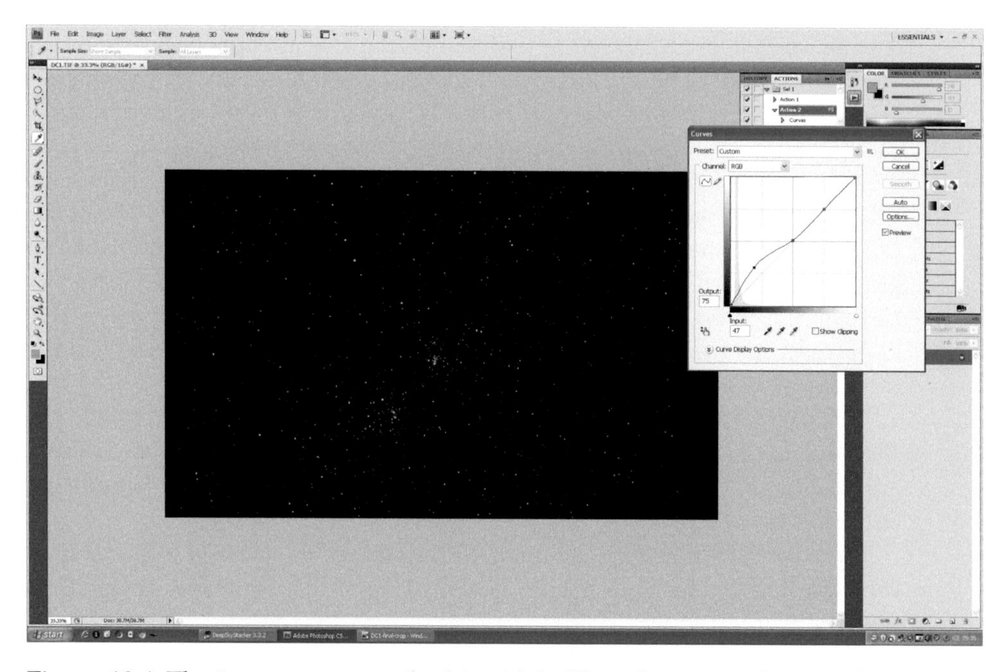

Figure 19.4 The image was stretched in *Adobe Photoshop* using four applications of the Curves function

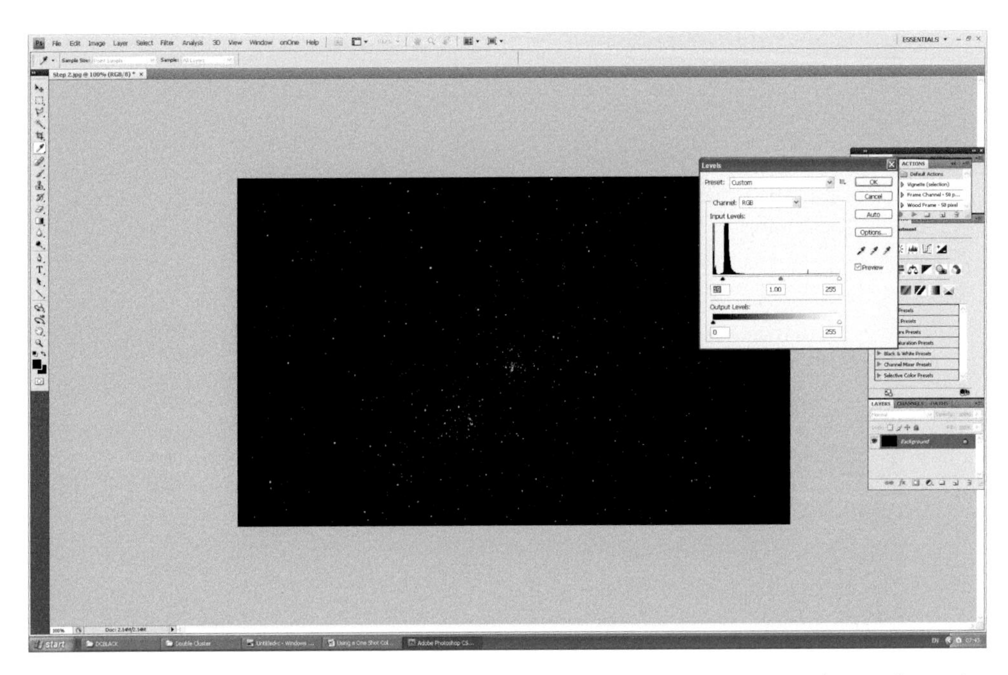

Figure 19.5 Using the Levels function, the black point was increased to reduce the background noise level

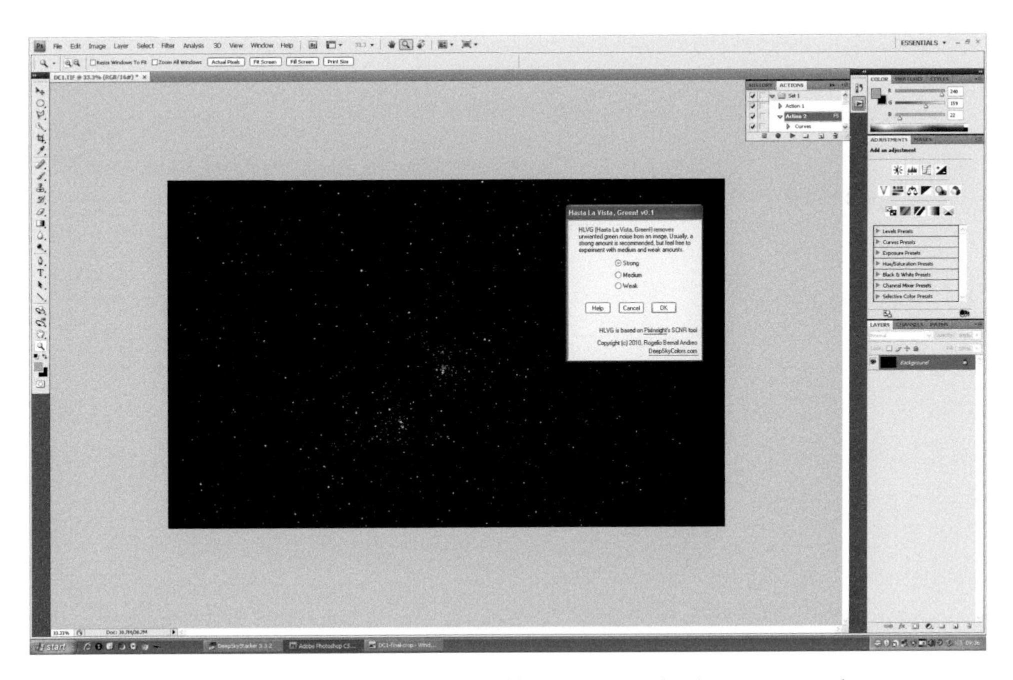

Figure 19.6 The 'Hasta La Vista, Green!' filter was applied to remove the green cast shown by the brighter stars

Figure 19.7 The Perseus Double Cluster imaged using a QHY8L colour CCD camera and William Optics 72 mm Megrez refractor

The final image of the Double Cluster, shown in Figure 19.7, was cropped to show the two clusters better. A number of red giant stars (which are actually orange) show up well.

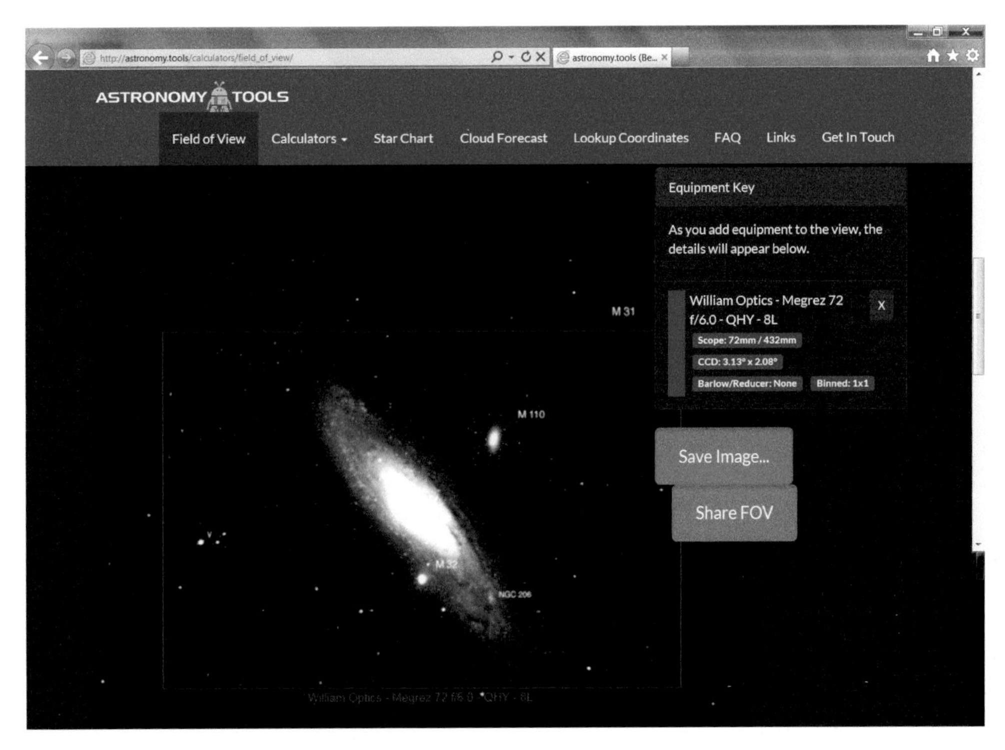

Figure 19.8 The field of view surrounding M31 obtained with a 72 mm refractor and QHY8L colour CCD camera

Imaging M31, the Andromeda Galaxy

Cooled CCD cameras are particularly appropriate for imaging galaxies – which are not called 'faint fuzzies' for nothing. Even with the very brightest, the Andromeda Galaxy, it is difficult to achieve a pleasing colour image within a reasonable time using a DSLR camera.

An excellent tool to select an appropriate focal length telescope with which to image an object can be found at: http://astronomy.tools/calculators/field_of_view. Having chosen the object, in this case M31, and QHY8L camera, telescopes of differing focal lengths can be selected to show how the object will fit into the field of view. With a field of view of 3.13 × 2.08 degrees, my 72 mm William Optics Megrez refractor proved to be a good choice (as seen in Figure 19.8) and this was used with a Teleskop Service 2 inch field flattener and Astronomik CLS light pollution filter.

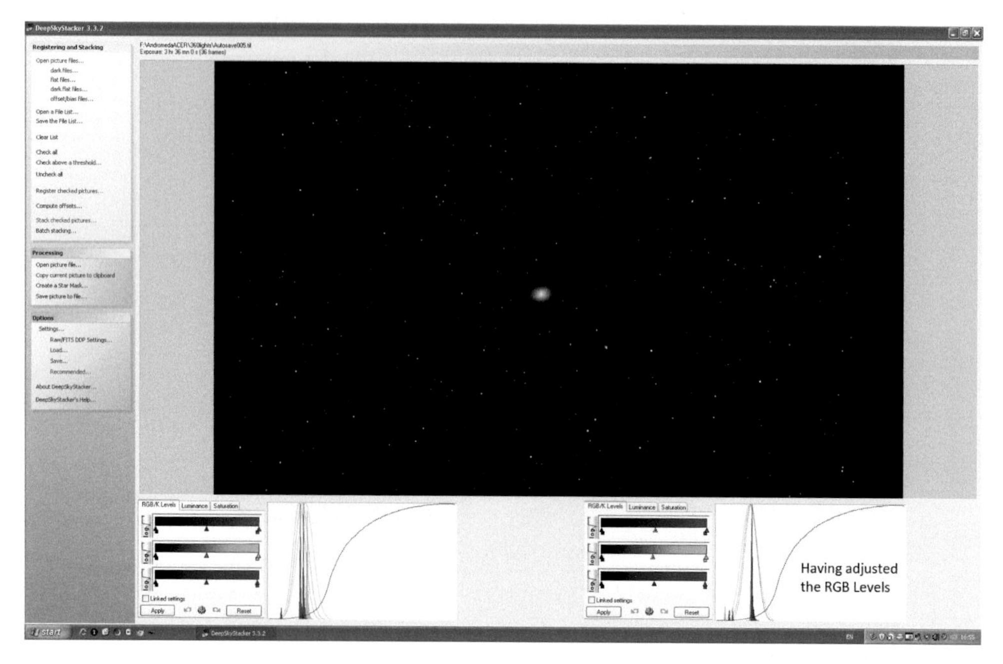

Figure 19.9 The output from *Deep Sky Stacker* with inset showing adjustment of the colour balance

The Astro-Physics Mach1 GTO mount was guided using the *PHD* software package utilising a Rother Valley Optics 80 mm, f/5 refractor (a superb guide scope) and QHY6 guide camera mounted in parallel to the main telescope. Using the 4×4 binning mode of the QHY8L to give high sensitivity, the galaxy was aligned along the major axis of the camera (though aligning along a diagonal might be more aesthetic). Then, by using the planner within the supplied imaging capture software package *EZcap*, a total of thirty-six 6-minute exposures were taken to give an effective exposure of 3 hours 36 minutes, with the individual frames being saved as FITS files. The 36 'light' FITS files were imported (with the appropriate 'GBRG' deBayer matrix selected) into *Deep Sky Stacker* for alignment and stacking. To these were added a set of 'dark' frames taken with the same exposure time and sensor temperature, which removes any hot pixels and amplifier glow. The result of stacking the 36 exposures is shown in Figure 19.9. To the lower left of the resultant image is a histogram of the three colours. The green histogram is well to the right of the red and blue histograms, indicating that the camera is more sensitive to the green. To correct the colour balance, the sliders of the green and red colours were adjusted to bring them into alignment with the blue, as seen in the Figure 19.9 inset.

The resultant image was stretched using several applications of the Curves function in *Adobe Photoshop*, with a final adjustment of the 'black point' within the Levels command to give the result shown in Figure 19.10, which was saved as 'M31Stars+nebula'.

Figure 19.10 The stretched image having used the 'Curves' and 'Levels' image adjustments in *Adobe Photoshop*

To allow the galaxy image to be enhanced without badly affecting the stars, the 'nebula' and 'stars' parts of the image were separated. The 'Dust and Scratches' filter in *Adobe Photoshop* was applied to the 'M31Stars+nebula' image with an amount of eight pixels to remove the majority of stars from the image and the remaining brighter ones were cloned out from adjacent areas. This image was then saved as 'M31Nebula', as seen in Figure 19.11.

The 'M31Stars+nebula' image was re-loaded, copied and pasted over the nebula image to give two layers so that, when the two were flattened with the 'Difference' blending mode selected, an image containing the stars alone was produced and saved as 'M31Stars' as seen in Figure 19.12. It was encouraging to see colour in individual stars.

Local contrast enhancement was then applied to the nebula image using the 'Unsharp Mask' filter with a radius of 250 pixels and an amount of 25 per cent (selected by trial and error) to give the result shown in Figure 19.13.

The 'M31Stars' image was reloaded, copied and pasted over the enhanced 'M31Nebula' image so that when the two layers were flattened with the 'Screen' blending mode selected, the stars were added back to the Nebula image to give the final M31 image shown in Figure 19.14. One is able to control how prominent the stars appear relative to the galaxy by adjusting the opacity of the stars layer and a value of 90 per cent was chosen to reduce their brightness somewhat.

Figure 19.11 The 'Nebula' image, having removed the stars

Figure 19.12 The 'Stars' image obtained by differencing the images of Figures 19.10 and 19.11

Figure 19.13 The enhanced 'Nebula' image

Figure 19.14 M31, the Andromeda Galaxy, imaged with a 72 mm refractor and QHY QHY8L one shot colour camera with 3 hours, 36 minutes total exposure time

The resolution is limited by the use of a small aperture refractor and the fainter parts of the galaxy would be better seen with a significantly longer total exposure time, but it does show that a pleasing image can be achieved within a reasonable imaging period. No doubt a similar result could be obtained if using a DSLR, but with shorter exposures, perhaps of 1 minute, to lessen the dark current within the pixel wells. I suspect that a total exposure time of 8 hours might well be needed to reduce the noise levels to the commendably low level achieved in the QHY8L image.

20
Cooled Monochrome CCD Cameras

The latest DSLR cameras make a very good job of astroimaging and can, of course, be used for general photography as well, so why go to the expense of buying a cooled CCD camera? The reasoning was outlined in Chapter 19. All imaging chips produce dark current which increases with exposure time and is also highly dependent on the temperature of the chip; that of a typical chip drops by half for each drop of 6° in temperature. So, by cooling the chip to 30° Celsius below ambient temperature, the sensor temperature, which would otherwise stabilise at about 12° Celsius above ambient temperature, will be at a temperature some 42 degrees lower and the dark current will have dropped by about 128 times, so allowing longer exposures to be taken before the dark current (which fills up the pixel wells and reduces the dynamic range of the sensor) becomes a problem. A second advantage is that the dark current noise that will be present in the image will have reduced by a factor of more than ~11, so images will look smoother. Given dark skies that do not suffer from light pollution, this can allow images to reveal faint nebulosity that would otherwise be lost in the noise. When significant light pollution is present, the exposure times, and hence the dark current contribution, have to be less before the skylight becomes obtrusive and so cooling does not confer such a great advantage. The latest chips have very low dark currents and it is rarely worth cooling them to below about –20° Celsius. This temperature can normally be reached with the single stage Peltier cooling employed in the CCD cameras aimed at the amateur market.

Aspects of Cooled CCD Cameras

Chip Size

The dimensions of the CCD chip allied to the telescope's focal length dictate the field of view (FOV) of the resulting images. The chip dimensions in millimetres divided by the focal length in millimetres gives the field of view in radians. Multiplying this by 57.3 gives the field of view in degrees. So, obviously, a bigger chip will provide a bigger field of view. For example, my 80 mm refractor has a focal length of 550 mm and if this is used with a CCD camera whose sensor has dimensions of 18 × 13 mm, a field of view of 1.9 × 1.4 degrees results. For those who have been used to using APSC sized sensors, this sensor size, which is that of the Kodak KAF-8300 chip used in many CCD cameras, is probably the smallest that one might wish to purchase.

However, the biggest chips can highlight telescope problems. If the telescope does not have a sufficiently large flat field, star images will become blurred towards the edge and the image may also suffer from vignetting (darkening towards the corners). As described in Chapter 8, it is possible to purchase field flatteners to overcome the former problem, such as the Teleskop Service 2 inch field flattener, whilst the latter problem can be corrected by the use of 'flat frames'. The camera supporting a larger chip is likely to be heavier, so will need to be very well supported to ensure that the CCD chip remains perfectly at right angles to the optical axis of the telescope. It is thus important that the focuser can handle the weight, and often telescopes can be bought with upgraded focusers specifically for this reason.

Pixel Size

It is important that a CCD chip has an appropriate pixel size to match the focal length of the telescope and the expected seeing conditions at your observing site. But it is worth pointing out that pixels can be binned to increase their effective area (and hence sensitivity). In most imaging software pixels can be binned as 2×2 (reducing the resolution by a factor of two but increasing the sensitivity by a factor of 4), 3×3 (3 and 9) and 4×4 (4 and 16). As we will see, this can allow a chip with small pixels to be optimised for different scopes and conditions. Binning is often employed when focusing the camera and aligning on the object to be imaged, as shorter exposures can then be used to speed up the process.

The combination of pixel size and focal length gives rise to what is termed the 'image scale' of the system, which is expressed in arc seconds per pixel. To calculate this, divide the pixel size (in microns) by the focal length (in millimetres) and multiply by 206.3. For example, when my Takahashi FS102, which has a focal length of 820 mm, is used with a CCD camera that has a pixel size of 5.4 microns, an image scale of 1.36 arc seconds per pixel results.

How does this relate to the resolution that the telescope will achieve? In most locations the effective resolution is limited by the seeing. This could well give typical stellar images that have 'full width at half maximum' (FWHM) of 3.5 or so arc seconds. The FWHM is the width of the star profile measured when the brightness has dropped to half its maximum value. One might thus think that an image scale of 1.36 arc seconds is too small. However, sampling theory (the Nyquist theorem) implies that the image scale be at least half, and preferably a third, of the apparent stellar size. So with seeing of 3.5 arc seconds this would indicate an image scale of ~1.2 arc seconds. This suggests that the FS102 + 5.4 micron pixel size is a pretty good combination unless the seeing happens to be particularly good. In fact, images will start to look blurred only when the image scale exceeds seven or so arc seconds and even an image scale of three arc seconds could produce images with quite reasonable detail.

Let us consider the image scale if the scope in use is my Celestron 9.25 inch Schmidt–Cassegrain. It can be used with its native focal length of 2350 mm or, when used with the f/6.3 focal reducer/flattener, with a focal length of 1480 mm. The image scales are then 0.47 and 0.75 arc seconds respectively. In both cases, under typical seeing conditions, the image will be over sampled with a 5.4 micron pixel size sensor and one could certainly employ 2×2 binning (or even 3×3 when using the longer focal length) and so significantly reduce the required exposure times. Thus a camera with ~5 micron pixels and capable of being used in a binned mode is a pretty good choice. So, if the seeing conditions at your site are average, consider cameras that will give an image scale in the range 1.0–3.0 arc seconds per pixel. It should be pointed out that the fundamental resolution of the telescope has some input into the calculations, as no matter what the 'seeing' might be, the resolution will be limited by the aperture of the telescope. This will not usually affect the conclusion, but suppose one is using a 14 inch Celestron telescope (with 3910 mm focal length), having a theoretical resolution of ~0.4 arc seconds, under skies having exceptional seeing (Damian Peach's images of the planets taken in Barbados come to mind), then a far smaller imaging scale will be appropriate. Using a camera with a pixel size of 7.5 microns would give an image scale of 0.4 arc seconds. If a 2× Barlow is employed, this would halve to 0.2 arc seconds and so be capable of extracting exceptionally detailed images.

Anti-blooming Measures

CCD chips can have a problem if very bright stars are included within an image. The electron wells of pixels where the light falls can overflow and bleed into the surrounding pixels, most noticeably into a vertical or horizontal streak along a row of pixels. This effect is called 'blooming' and would, for example, seriously detract from an image of the Pleiades cluster when one is trying to detect the faint nebulosity surrounding the very bright stars. Blooming affects CCD chips that have what are called 'non anti-blooming gates' (NABG). (Interestingly, this effect can be used to focus an NABG CCD camera. Simply observe a bright star and adjust the focus until the streak is at its longest and thinnest. This means that the star's light is largely concentrated onto one or two pixels and will then be in perfect focus.)

CCD chips can also be constructed with 'anti-blooming gates' (ABG). In this case the electron well is surrounded by circuitry that bleeds off the excess electrons before they can spill into adjacent pixels. Such chips would thus seem to be the obvious choice; however, this circuitry takes up space, so the effective collecting area of each pixel will be less and hence the quantum efficiency of the chip will be lower and so longer exposures would be required. There used to be quite a big difference in quantum efficiency and so for some applications NAGB chips would be the preferred choice. More recently, CCD chips have become available that use micro lenses above each pixel to concentrate the light from the whole pixel area into the electron well. This naturally increases the quantum efficiency, which may well have a peak of 56 per cent, as opposed to an NAGB peak efficiency of ~80 per cent. Not as good but, as the difference is not now so marked, a chip utilising micro lenses may well be the best choice.

Dynamic Range

CCD cameras will usually employ a 16-bit analogue to digital converter (ADC) in the chip readout electronics, as opposed to 12- or 14-bit converters in DSLR cameras. This too can make a slight improvement in image quality. However, the number of bits is not quite as important as one might think. Many images are now made by stacking a number of sub-exposures and not only does this have the effect of reducing the noise in the image, it also increases the effective number of bits in the conversion process.

The Narrowband Advantage

Under light polluted skies, astroimagers are now tending to image with narrowband filters. In this case, colour CCD cameras, such as the QHY8L described in Chapter 19, are at a major disadvantage when imaging the light from particular emission lines such as, for example, the H-alpha line at 6563 angstroms. Often, imagers will add an H-alpha layer into the overall image and it can provide added contrast and definition and highlight the star formation regions. It is possible to use a narrowband filter with a colour CCD camera, but in the case of H-alpha only one in four pixels will be utilised, reducing both the resolution by a factor of two and the sensitivity by a factor of four. So, for the many of us who live under light polluted skies, a mono CCD camera is probably the most versatile. But it has to be said that the overall cost will be significantly higher and my advice would be to first acquire a cooled colour camera as described in Chapter 19, and perhaps then move on to a monochrome camera – particularly if one desires to carry out narrowband imaging.

Choosing a CCD Camera

To some extent it depends on what you wish to use it for, but in general those with larger CCD chips and more pixels will be more versatile. Small pixels will give a good imaging scale when used with short focus refractors for wide field imaging and their pixels can be binned for use with long focal length telescopes. For general use an AGB chip that eliminates blooming is a sensible choice, and if it employs micro lenses, not that much less sensitive.

Figure 20.1 The Atik Titan (left) and QHY QHY6 (right). Each has USB2 and ST-4 guiding ports along with a 12 volt power input for Peltier cooling when required

A First Cooled CCD Camera

It is now possible to buy a cooled monochrome CCD camera for a reasonable sum of money. The Atik Titan and the QHY QHY6, seen in Figure 20.1, are two examples. As one might expect, the sensor sizes are not large (of the order of 6 × 5 mm). The Atik Titan employs a 325 K, 659 × 494 array of 7.4 micron square pixels, while the QHY6 uses a 752 × 582 array giving 440 K pixels of 6.5 × 6.25 microns. The other cost saving aspect is that, though the sensor is cooled with a single stage Peltier cooler and can reach 25 to 30° Celsius below ambient temperature, there is no temperature control system to keep the sensor at a specific temperature and no way of knowing what that temperature is. This means that dark frames, if used, need be taken at the same time as the light frames so that the sensor temperature does not change significantly between them. However, the cooling will both significantly reduce the number of hot pixels present and also reduce the dark current, so is well worth having.

Should one wish to move on to a CCD camera with a larger sensor, such as those described below, then this 'first CCD camera' may well not be wasted. They make excellent guide cameras and both those mentioned above are equipped with the standard ST4 guide port to allow them to be easily used in this role, as described in Chapter 9 and Appendix D.

Large Sensor Cooled CCD Cameras

The problem with cooled CCD cameras, particularly those with large chips for use in wide field imaging is that, in the past, they have been very expensive. More recently, a Kodak KAF-8300 CCD chip has become available in both mono and colour versions. This utilises 8.3 million, 5.4 square micron pixels in a 3326 × 2504 array across an 18 × 13.5 mm chip. Anti-blooming gates (which would tend to reduce its sensitivity) and micro lenses (which increase it) are used and it has a peak quantum efficiency of around 56 per cent in its mono version.

This chip has been incorporated into a number of CCD cameras, for example the SBIG STF-8300 in mono and colour versions, the Apogee Alta® F8300 camera in

Figure 20.2 The SBIG STF-8300 and Filter Wheel and Atik 838L+ monochrome CCD cameras

mono and colour versions, the QSI 583 mono camera (which can include an integral filter wheel), the Moravian Instruments G2 8300 camera incorporating an integral filter wheel, the Atik 383L+ mono camera and the QHY QHY9 in mono and colour versions. Celestron have introduced their 'Nightscape' colour CCD camera using the colour version of the same CCD sensor. They all represent excellent value for money and have made large size, multi-megapixel, imaging much more accessible.

It is very likely that when using large sensor CCD cameras, the problem of field curvature may well become a problem, with star images at the corners of the field becoming blurred and distorted. As discussed in Chapter 8, a field flattener (often also acting as a focal reducer) will almost certainly be used, unless the telescope already incorporates correcting optics in its light path, as in the Celestron 'Edge HD' telescopes and the Vixen VC200L.

Monochrome CCD Imaging

Before going to the expense of buying a full set of filters and a filter wheel it is not a bad idea to try some monochrome imaging. In the same way that monochrome photographic images can have their own impact, so can monochrome astrophotographs. In particular, faint galaxies show rather little colour and better images might well result if the full imaging time available is spent gaining a monochrome 'luminance' image.

A Monochrome Image of the Leo Triplet

As mentioned in Chapter 8, a far less expensive alternative to a short focal ratio Newtonian astrograph is to use an 8 inch, f/4 Newtonian combined with a coma corrector.

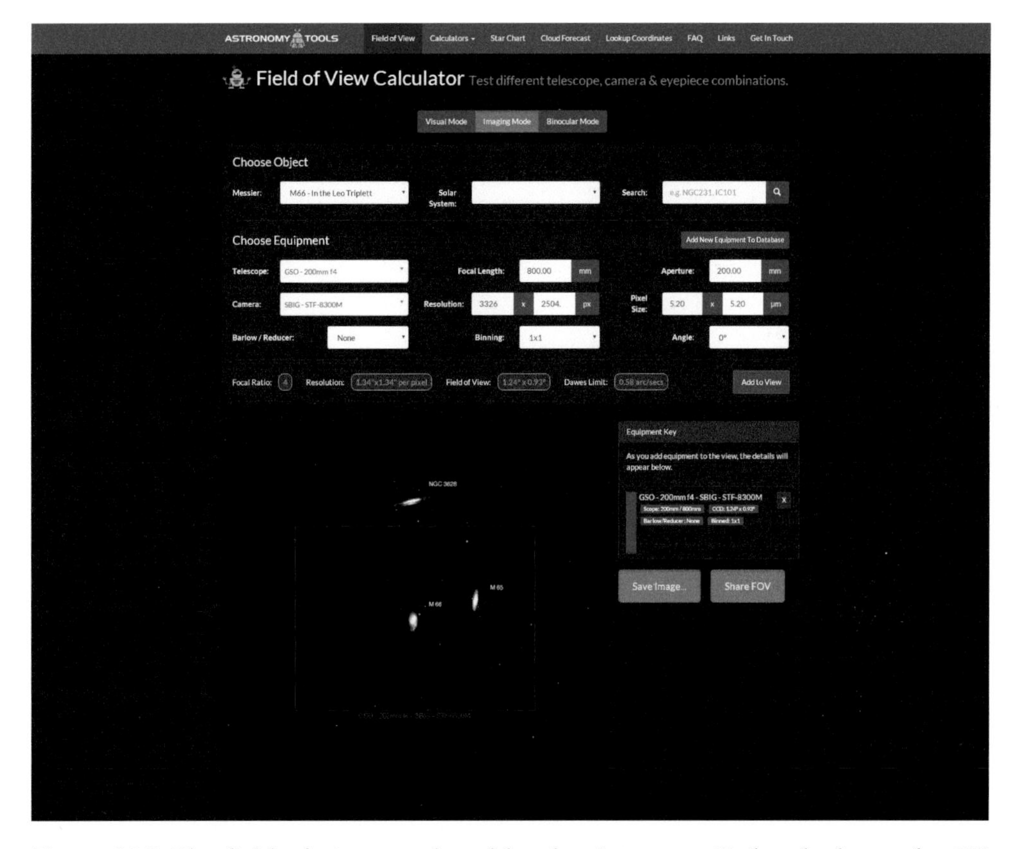

Figure 20.3 The field of view produced by the *Astronomy Tools* calculator of a 200 mm, f/4 Newtonian and CCD camera using the Kodak KAF-8300 CCD

Some of these are specifically designed to have a large secondary mirror in order to reduce vignetting at the corners of the image. I do not have one, but I do have a Meade 8 inch (200 mm), f/4 Schmidt–Newtonian which uses a spherical mirror allied to a corrector plate on which is mounted the secondary mirror. There will therefore be no diffraction spikes on bright stars. The coma is said to be roughly half that found in a Newtonian and they are said to work well with a Baader Coma Corrector.

Using the *Astronomy Tools* field of view calculator, as shown in Figure 20.3, I discovered that the Leo Triplet would be nicely encompassed using the Meade telescope and SBIG ST-8300M camera[1]. The field is centred on M66, but it can be seen that by adjusting the telescope pointing, all three bright galaxies would be included. The Leo Triplet (also known as the M66 Group) is a small group of galaxies about 35 million light years away in the constellation Leo and consists of the spiral galaxies M65, M66 and NGC 3628, also known as the Hamburger Galaxy. NGC 3628, discovered by William Herschel in 1784, has a broad and obscuring band of dust located along the outer edge of its spiral arms.

[1] The SBIG ST-8300M camera is an earlier version of the SBIG STF-8300M.

Figure 20.4 A light frame, left, showing vignetting towards the corners with, right, the 'Master Flat' produced by *Deep Sky Stacker*

The Astro-Physics Mach1 GTO mount used for the image capture was guided using the *PHD* guiding software package and a Rother Valley Optics 80 mm, f/5 refractor (a superb guide scope) and QHY QHY6 guide camera mounted in parallel to the main telescope. To increase the camera's sensitivity by nine times I chose to use its 3×3 binning mode. This gave an image size of 1117 by 844 pixels rather than 3352 by 2532, so, of course, the resolution was reduced by a factor of three. The galaxies were aligned within the frame, and the sensor was cooled down to −20° Celsius. A series of thirty-six 120-second exposures (light frames) were taken, followed by 10 dark frames having the same exposure and sensor temperature. As the secondary is not over sized, the images showed significant vignetting towards the corners. The effects of two dust motes are also just visible. As I had expected, this was one imaging example where 'flat' frames were required, so the following morning I pointed the telescope up towards a cloudy sky and placed a translucent plastic sheet over the corrector plate. It is important that the camera orientation is the same as when the light frames were taken and it is recommended that the exposures are set to give a peak brightness of ~75 per cent. A highly stretched single light frame is shown Figure 20.4, left, to show (somewhat overemphasised) the vignetting that was present.

A total of 20 flat frames were taken and these were imported into *Deep Sky Stacker* along with the light and dark frames. As part of the processing, *Deep Sky Stacker* produces a 'Master Flat', which is shown (again overemphasised) in Figure 20.4, right. The two dust motes are very obvious. Having changed the mode from monochrome to RGB and using the 'Eyedropper' tool, having opened the 'Info' window, one can explore the brightness across the image. The peak brightness was 144 (out of 256), with the four corners dropping to 106, 105, 126 and 122. This shows the importance of using the same camera alignment as when the light frames were taken!

Once calibrated, aligned and stacked, the TIFF output from *Deep Sky Stacker* was imported into *Photoshop*. The image was stretched with repeated use of the Levels command with the centre slider set to 1.20 – this having been made an 'Action'. After each every applications of the action, a Level adjustment was made to bring the black point (left slider) up to the point where the histogram rises. The use of flat frames had removed both the vignetting and the effects of the two dust motes, but the image showed a varying amount of light pollution across the image. To remove this, the standard technique was used: the image was duplicated, the 'Dust and Scratches' filter (8 pixels amount) was used to remove the stars, and the one bright star and three galaxies cloned out from surrounding areas before flattening the two layers using the 'Difference Mode'.

Some final editing 'tweaks' were then made to the image: the outer parts of the galaxies were noisy and so these were selected using the 'Polygonal Lasso' tool and a Gaussian Blur with a radius of one pixel applied, the three galaxies were selected and the 'Smart Sharpen' filter used to enhance them, and finally, the bright central star which appeared rather 'hairy' was smoothed by the use of some Gaussian Blur. The final result is shown in Figure 20.5.

An H-alpha Image of the North America and Pelican Nebulae

Monochrome images do not necessarily need to be made using full spectrum light, and given a 2 inch H-alpha filter, it is possible to make very attractive images in the light of the hydrogen emission line in the deep red. An excellent H-alpha imaging target is the region containing the North America and Pelican Nebula that lies close to the star Deneb in Cygnus. A colour image taken with a modified DSLR camera was included in Chapter 11. A William Optics 72 mm Megrez refractor equipped with a Teleskop Service 2 inch field flattener was coupled to my SBIG ST-8300 camera (operating at −20° Celsius) with a 2 inch Baader H-alpha narrow band filter screwed to the 2 inch camera coupling that fits within the focuser. The telescope and camera were mounted on an unguided iOptron Mini-Tower, alt/az mount and 20 exposures of 120 seconds, each were taken whilst the region was high in the west, where field rotation was not too much of a problem. All frames were saved as 16-bit TIFF files.

As the nebulosity is very faint, to increase the sensitivity by a factor of four (but reducing the resolution by a factor of two) the sensor's output was binned 2×2 and a further 20 dark frames were taken with the same exposure time and sensor temperature for use when the frames were calibrated and stacked in *Deep Sky Stacker*. In *Photoshop*, the resulting TIFF image was processed using curves, Levels and some local contrast enhancement. Even when using an H-alpha filter, many of the star images (as shown in Figure 20.6, left) are somewhat bloated and these detract from the nebula regions. This is where the 'Minimum' filter in *Photoshop* (Filter > Other > Minimum) can be useful. If this is applied directly to the whole image, the fainter stars will tend to disappear and the effect, even when using a single pixel radius, can be too severe. To allow greater control over the filter, the image should be doubled in size before it

Figure 20.5 An image of the Leo Triplet made using an 1 inch (200 mm), f/4 Schmidt–Newtonian, Baader Coma Corrector and SBIG ST-8300M camera with 72 minutes total exposure

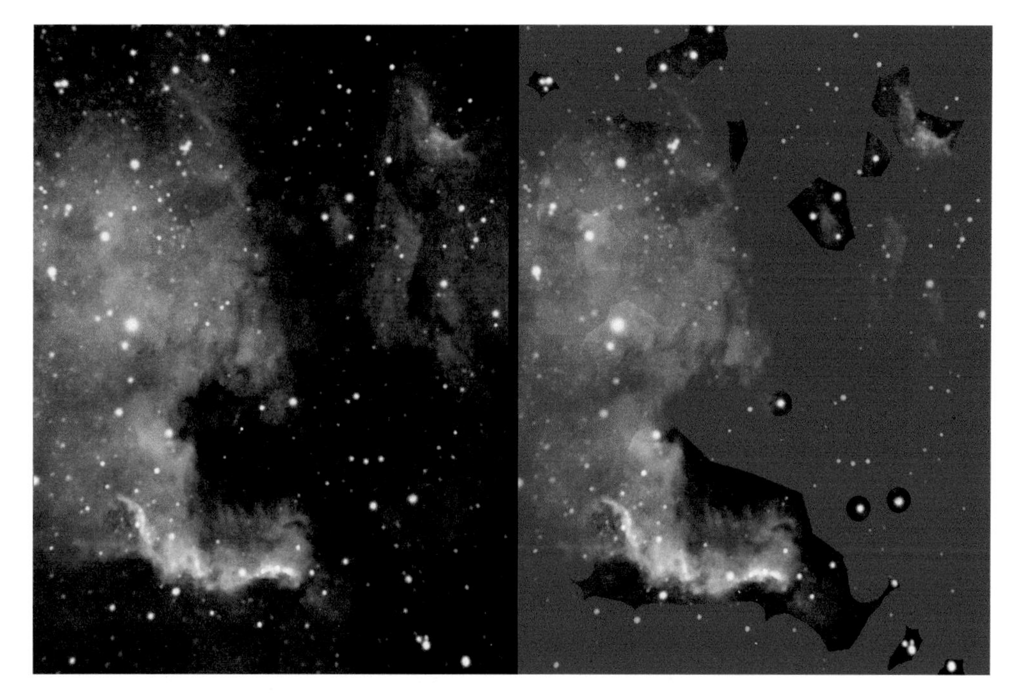

Figure 20.6 A mask to select only the brightest stars in the image

is applied, so a one pixel radius effectively becomes half a pixel. To avoid removing the fainter stars a mask was made to cover the regions including the brightest stars but avoiding the fainter stars within the image (shown in red in Figure 20.6, right). Having then applied the minimum filter, a more attractive image resulted, as shown in Figure 20.7, after having also applied some local contrast enhancement to the Pelican Nebula at upper right and to the lower part of the North America Nebula. The Pelican Nebula and the arc towards the bottom of the image, which is called the Cygnus Wall, are nicely shown.

Figure 20.7 An H-alpha image of the North America and Pelican Nebulae taken with a William Optics Megrez 72 mm refractor, Teleskop Service field flattener and SBIG ST-8300 CCD camera

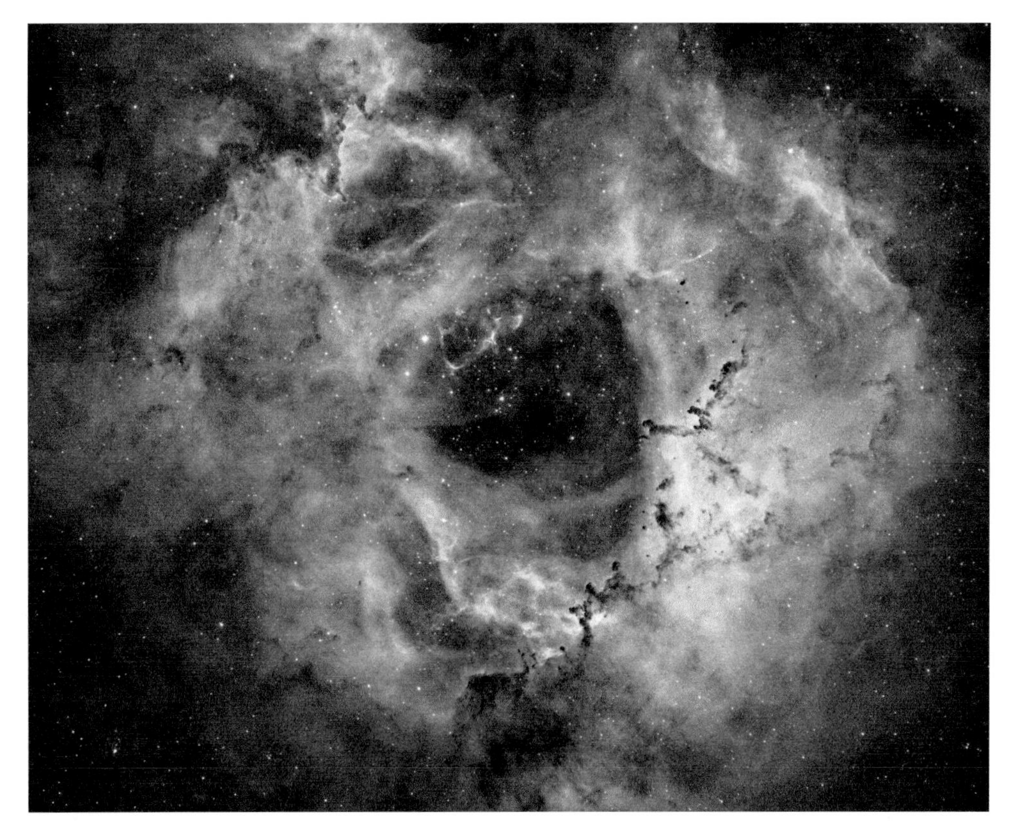

Figure 20.8 An H-alpha image of the Rosette Nebula taken by Christopher Heapy using a TeleVue NP 127is refractor, H-alpha filter and Atik 490ex monochrome CCD camera

An H-alpha Image of the Rosette Nebula in Monoceros

Figure 20.8 shows a beautiful H-alpha image taken by my colleague Christopher Heapy using a TeleVue NP 127is astrograph of the open cluster NGC 244 and the surrounding H-alpha nebulosity making up the Rosette Nebula in Monoceros. The 127 mm aperture refractor he used has a greater resolution than my 72 mm Megrez and the narrower band H-alpha filter that was used also helped to give greater contrast. I believe that this image does support the view that monochrome images can have real impact!

21 LRGB Colour Imaging

LRGB (luminance RGB) is the standard technique used by deep sky imagers to produce their images and involves taking a monochrome image – called the luminance image – to which colour is added from an RGB image which is constructed from three individual monochrome images taken through red, green and blue filters. An infrared cut-off filter (the L filter) is used when forming the luminance image to avoid any out of focus halos. Dichroic RGB filters, discussed below, also remove the infrared. Given the same total imaging time, the LRGB technique tends to give better overall results than simply taking R, G and B images. This is because the colour information in an image does not require such high resolution and the RGB images are often taken using 2×2 binning to increase the sensitivity by a factor of four, hence requiring less exposure time. So it is usually suggested that half of the total imaging time be spent taking the luminance image, with the remainder of the time split between the three binned colour images. Should the sensitivity of the camera be less in one wavelength band (often the red), the RGB times may be adjusted to aim to give equal quality images in all three colours.

LRGB Filters and Filter Wheels

When a monochrome camera is used, a filter wheel (not absolutely necessary but highly convenient) is used to select the colour being imaged with a monochrome CCD camera. The filter sets available until relatively recently used glass doped to transmit a particular colour range. Their bands overlap somewhat and the transmission efficiency is perhaps in the order of 70–80 per cent. Filter sets are now available, such as those manufactured by Baader Planetarium, which use dichroic filters that offer many advantages. Their individual passbands cut off very steeply and their transmission efficiency in band is well over 90 per cent. Only their blue filter is designed to pass the H-beta spectral line, so giving it the correct blue colour, but the O III line is passed by both blue and green filters so giving it its correct colour when the RGB image is produced (blue-green or teal), while also increasing the overall sensitivity at that wavelength. The red filter accepts both the (highly important) H-alpha and

Figure 21.1 The author's SBIG ST-8300M used for monochrome imaging and coupled to the SBIG CFW9 filter wheel. This is opened to show the R,G,B, and H-alpha dichroic filters along with an infrared cut-off filter through which to take the luminance images. The filter wheel is coupled to the camera through a 9-pin, D type connector

S II lines with near 100 per cent transmission, but then efficiently cuts off the infrared. There is a narrow gap at ~5800 angstroms between the green and red passbands to help suppress the emission from sodium street lamps. The L (luminance) filter simply cuts out the infrared emission. All three colour filters have equal sensitivity so that the same exposures can be made for each, which greatly simplifies automated imaging (assuming that the camera's sensitivity is fairly uniform across all three bands, which may not, in fact, be the case). Tests can be done imaging a white test card to find the relative exposures required in the three bands to give a white image when the RGB images are combined. Very similar LRGB filter sets are manufactured by Astronomik with their 'LRGB type 2 filters' and Astroden with their 'True Balance Filters'. Given the expense of the CCD camera and filter wheel it is simply not worth using lower cost filters than these, even though they are somewhat expensive.

With a typical five-slot filter wheel, the LRGB set could be supplemented with an H-alpha filter to add an H-alpha layer to the image. With seven-slot filter wheels both

S II (deep red) and O III (blue-green) narrowband filters can be included to allow for both LRGB and narrowband imaging (discussed in Chapter 22) to take place easily.

Filter wheels can be operated manually, or controlled directly from the imaging computer or via the CCD camera to which it is coupled. The last approach helps to reduce the number of cables between the computer and telescope and can also easily allow for filter changes in imaging sequences to be directly controlled by the imaging program. Many CCD cameras will have dedicated filter wheels which could have five or seven slots to hold either 1.25 inch or 50 mm diameter filters. The larger (and more expensive) filters are required when using large (such as APSC size) CCD sensors and even the somewhat smaller 8.3 Mpixel chip used in my SBIG ST-8300 camera when used with short focal length telescopes. I am using the SBIG CFW9 filter wheel in conjunction with the ST-8300 camera. This incorporates 1.25 inch mounted filters which are sufficiently large in diameter to give un-vignetted images when used with my f/8, 102 mm refractor, but when used with my f/6.8, 80 mm refractor, I have to accept some minor vignetting towards the corners of the frame, and these filters could not be sensibly used with my 200 mm, f/4 Schmidt–Newtonian. However, minor vignetting can be easily corrected for by taking 'flats' in the imaging process or, later, in post processing. SBIG now provide a filter wheel that takes a new size of filter that is 36 mm in diameter. This provides un-vignetted imaging with the ST-8300 series with any focal ratio telescope. The new filter size is cheaper than the 50 mm diameter filters that would otherwise have been needed.

In order to use a 200 mm, f/4 Newtonian and short focal length telescopes for imaging, I have acquired a set of 50 mm filters but not, as yet, purchased a filter wheel. As one is using each filter for quite long imaging periods it is not too much trouble to screw each filter into the 2" barrel that fits into the focuser and refocus the image before imaging with each filter. My current feeling is that I will not go to the expense of purchasing one.

As an example to illustrate this chapter I remotely imaged the galaxy M33, in Triangulum, using an ASA 8 inch Newtonian Astrograph situated at Nerpio in Spain at a height of 1680 ft. It was equipped with a monochrome Atik 383L+ 8.3 megapixel camera (similar to my own SBIG ST-8300) and an Atik EFW2 filter wheel incorporating dichroic Astronomik LRGB filters. The system has a field of view of 117 × 80 arc minutes. Luminance and RGB images were taken around new Moon with a total observing time of 10,800 seconds (180 minutes). Ten sub-frames, each of 300 seconds, were taken for the luminance image, eight sub-frames of 300 seconds for the green and nine sub-frames of 300 seconds for both the red and blue, making a total of 36 sub-frames. In addition, twenty 300-second dark frames were taken along with 15 flats for each of the LRGB filters – making a total of 136 FITS images. If one wishes to inspect individual FITS frames, they can be imported into *IRIS*, which will immediately show a greyscale image. I used *Deep Sky Stacker* to calibrate, align and stack the four LRGB images. For each image the light frames with their appropriate flats and dark frames were loaded into *Deep Sky Stacker*. (As the dark frames had the same exposure time as the light frames, bias frames were not required.) Each calibrated and stacked image was then able to be exported as a TIFF file for further processing in *Photoshop*.

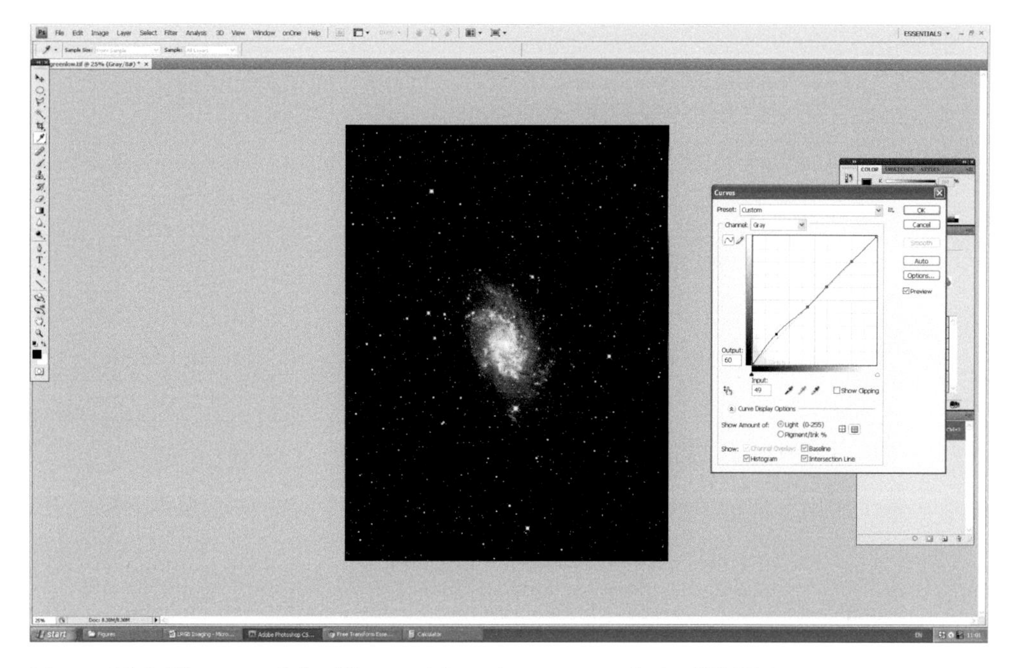

Figure 21.2 The use of the 'Curves' function to stretch the LRGB images

The first step was to stretch each image. As shown in Figure 21.2, this is done using seven applications of a very gentle curve, followed by the setting of the dark point to the point where the histogram rises sharply, with the use of Levels.

At this point all four monochrome images looked broadly similar. The next step is to produce a RGB colour image. All three monochrome images are loaded into *Photoshop* and a 'new' image opened up with the same dimensions and in RGB 16-bit colour mode, as seen in Figure 21.3. (The appropriate image size may be set automatically, but, if so, it will be in greyscale mode and will need to be changed to RGB, 16-bit, colour mode.)

The 'Channels' box is opened up (Window > Channels) and shows three blank red, green and blue frames as well as the RGB channel. Each monochrome image is selected (Ctrl-A), copied (Ctrl-C) and pasted (Ctrl-V) into the window in the appropriate channel box and then a colour image appears, as seen in Figure 21.4.

Unless the three RGB images are perfectly aligned it may well look very messy: in this case the blue image was offset up and to the right of the green and red images. The three images will then have to be aligned vertically and horizontally and, perhaps, rotated slightly. To achieve this, within the channels window, the 'eye' within the red channel box is checked, so removing the eye and the red channel from the image so that a green/blue composite image remains (Figure 21.5).

At the top left of the tools menu, the 'Move' tool is selected and now, using the cursor keys, the blue image can be aligned with the green image. Hopefully, the orientation of the camera will not have changed between the two images, but if it has, the blue image will also need to be rotated. This is done by selecting the blue channel

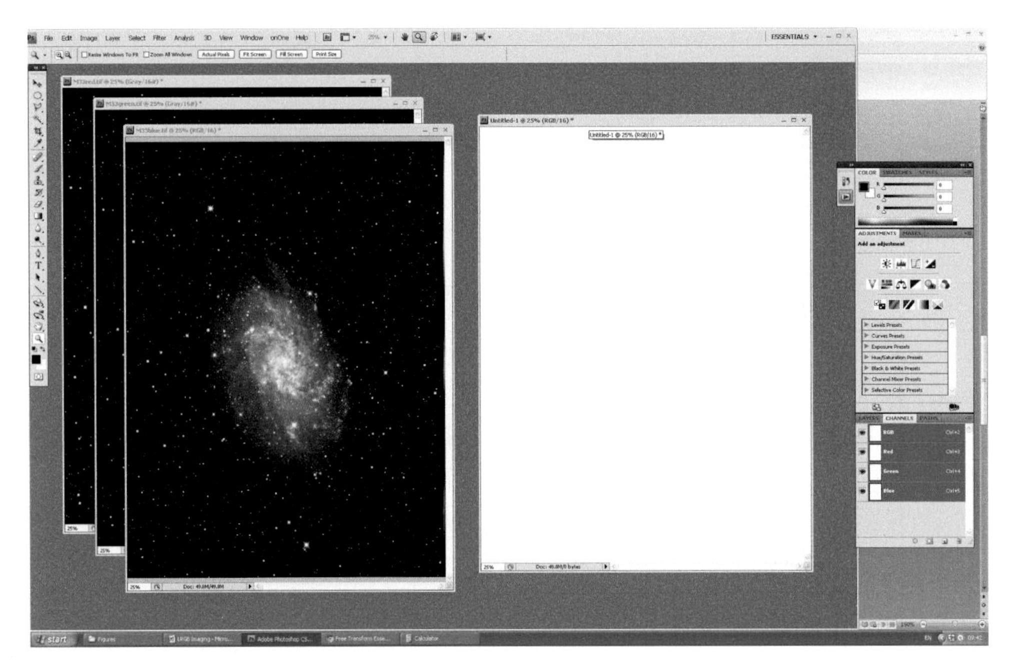

Figure 21.3 A new image is opened in which to add the individual R, G and B monochrome images and so produce a colour image

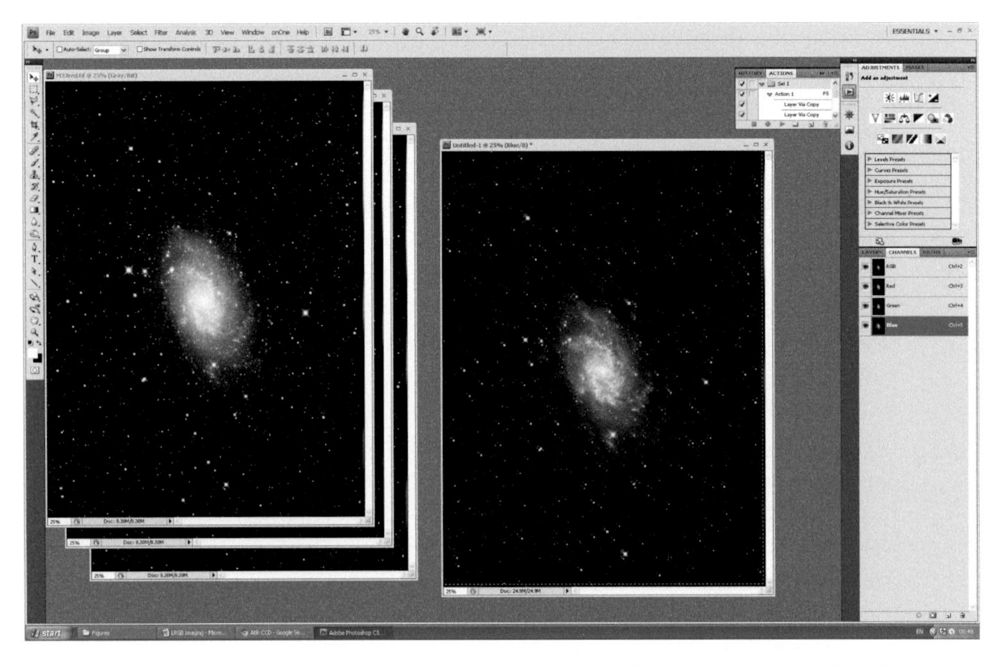

Figure 21.4 The initial production of the RGB colour image (in which the alignment is not correct)

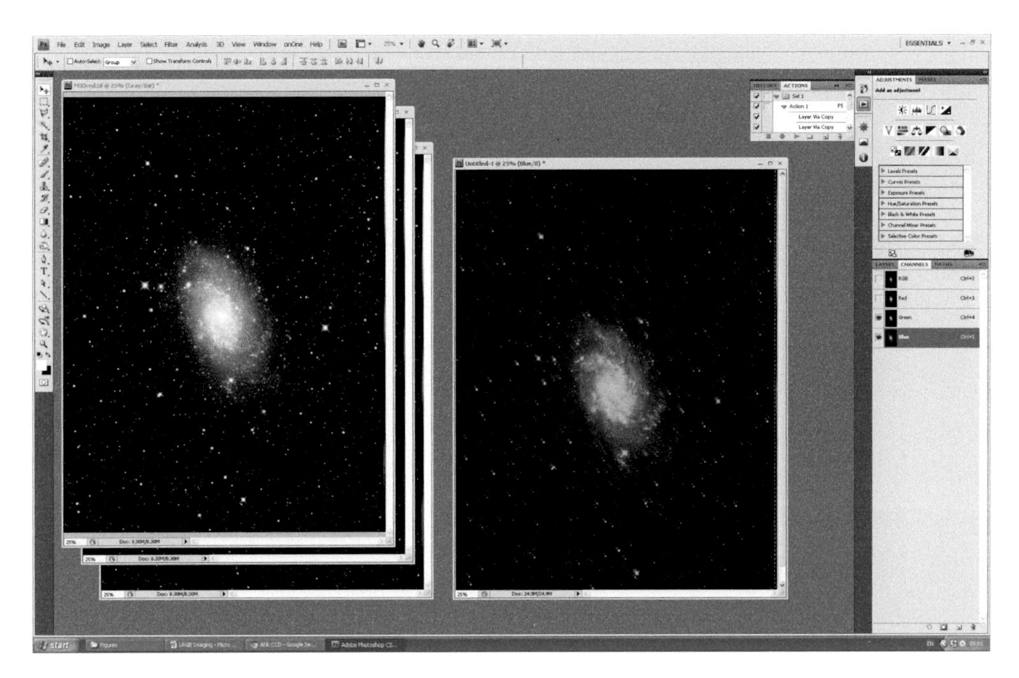

Figure 21.5 The blue/green composite image displayed to enable them to be aligned

using Ctrl-A and then initiating a transform using Ctrl-T. Small boxes will appear at the corners and sides of the image. If the cursor is placed *outside* the image, say outside the top right corner, a rotation symbol appears and, moving the cursor will rotate the blue image over the green one. Moving the cursor *within* the image then allows one to use the cursor keys to align the two images. This is a somewhat iterative process that may take some time! To finalise the transform of the blue channel, either the image is 'double clicked' or the 'tick' symbol that has appeared above the image is checked. To then align the red channel, the 'eye' within the blue channel box is checked (so removing it and just leaving the green image) and the eye within the red channel box is checked to give a red/green image. The red channel box is then checked and turns blue. This channel is then aligned over the green channel just as for the blue channel. With all three eyes checked, an aligned full colour image (Figure 21.6) appears and should, I think, be saved. The colour balance is not correct, with more blue and less red relative to the green within the image.

One can then spend some time adjusting this image to give the best looking result. There are several ways of doing this: one can select each channel in 'Levels' and adjust its black, mid and white points, or perhaps use the 'Colour balance' controls. Within a free, downloadable, filter set for *Adobe Photoshop* found at www.deepskycolors.com are two tools that might help. One, 'WhiteCal', allows one to select an area of the image (using the 'Rectangular Marquee' tool) that should be white and will automatically adjust the relative levels of the three colours to provide the correct ratios to

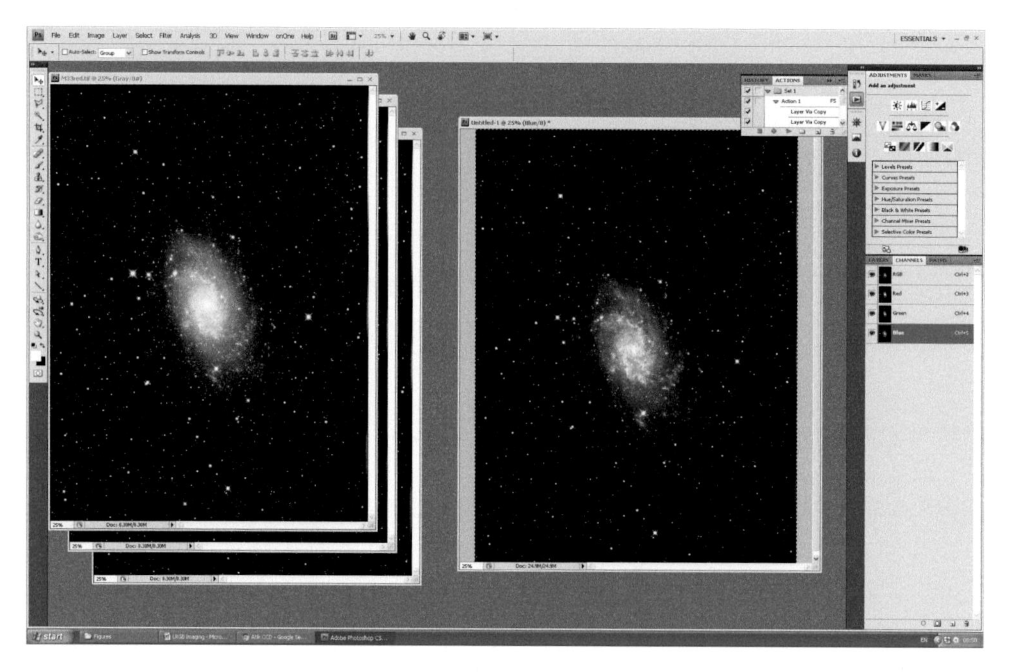

Figure 21.6 The aligned RGB image before smoothing and adjusting the colour balance

make this area white, while a second, 'RGBC', allows the relative sensitivity in the three colour bands to be set manually so that a correct colour balance is achieved.

This image may well be rather noisy as usually less time is spent imaging the R, G and B filtered images and, owing to the presence of the filters, the sensitivity of the CCD will be less. One does not need such a high resolution for the colour image as for the luminance image and so, to reduce the noise, one can apply a Gaussian Blur of a few pixels. An alternative smoothing method is to first reduce the image size by 50 per cent and then increase it by 200 per cent to return to the original size. *Photoshop* will have averaged over 4 pixels when it made the reduced image. Although not done for this particular image, if the RGB images were taken with 2×2 binning (giving a smaller but less noisy image), it must be increased by 200 per cent before being used to colour the luminance image. The saturation of this image was rather low and there are several techniques, outlined below, that can increase the colour in the image.

Enhancing the Overall Colour in an Image

Perhaps the most obvious method is to simply increase the saturation: Image > Adjustments > Hue/Saturation. Another technique is to use the Shadows/Highlights dialogue box: Image > Adjustments > Shadows/Highlights. This, by default, lifts up the darker parts with the 'Amount' slider (at the top) at 50 per cent, which is not wanted but, nevertheless, should be left there, as if the slider is returned to zero, no colour adjustment can be made. It may be necessary to tick the 'Show More Options' box to reveal the Colour Corrections slider. This can then be moved to the right to perhaps +50 to nicely enhance the colour. Having clicked OK, the Levels dialogue box is entered, and the middle and perhaps the left pointer (to set the black point) adjusted to bring the overall brightness back to what is desired.

Using the 'Lab Color'[1] Tool to Enhance the Colour in an Image

A technique that works very well is to convert the image into Lab Color: (Image > Mode > Lab Color). When adjusting the saturation in RGB mode, the brightness of each pixel has to adjust as well and this may well introduce some chrominance noise into the image. However, in Lab Color, the luminance contrast and colour are separated and so the saturation can be increased without altering the brightness. There are three channels: Lightness, 'a' and 'b'. The image is loaded, converted into lab Color and saved with a '_sat' suffix added to the file name (this is to make sure that an inadvertent key stroke will not replace your original image). The Channels window (Windows > Channels) is opened and the 'a' channel selected, which will appear grey with differing shades, depending on the image. The Levels command (Image > Adjustments > Levels) is entered and a histogram peaking at around the centre will be seen. Now the input levels are adjusted with the two outer sliders to give 55 and 200 (rather than 0 and 255). The 'b' channel is then selected and the same process carried out. Selecting the top (Lab) image in the Channels box will show a highly saturated image, which is converted back to RGB colour mode: Image > Mode > RGB. The original image is then opened and the saturated version copied (Ctrl-A, Ctrl-C) and pasted on top (Ctrl-V) to make a second layer. In the Layers window, the two images will be one above the other. Finally the Opacity slider is adjusted to give the required amount of saturation before the image is flattened (Layer > Flatten Image) and saved. This sounds like a rather involved process, but after a couple of tries it does not take too long and, I promise you, the result is far better than applying a simple saturation adjustment.

One problem with using these techniques is that it will increase the saturation of all areas of the image, even those that are well saturated already. *Photoshop CS4* and above and *Adobe Camera Raw* have a slider called 'Vibrance' (Image > Adjustments > Vibrance) that boosts only the saturation of lesser saturated colours, so it will not over saturate those colours that are already bright.

Using the 'Match Color'[2] Tool in *Photoshop* to Enhance the Colour of an Image

There is a very quick and easy way to enhance a colour image in later versions of *Photoshop* which I have found to give superb results. The image is opened and a

[1] American spelling.

[2] American spelling.

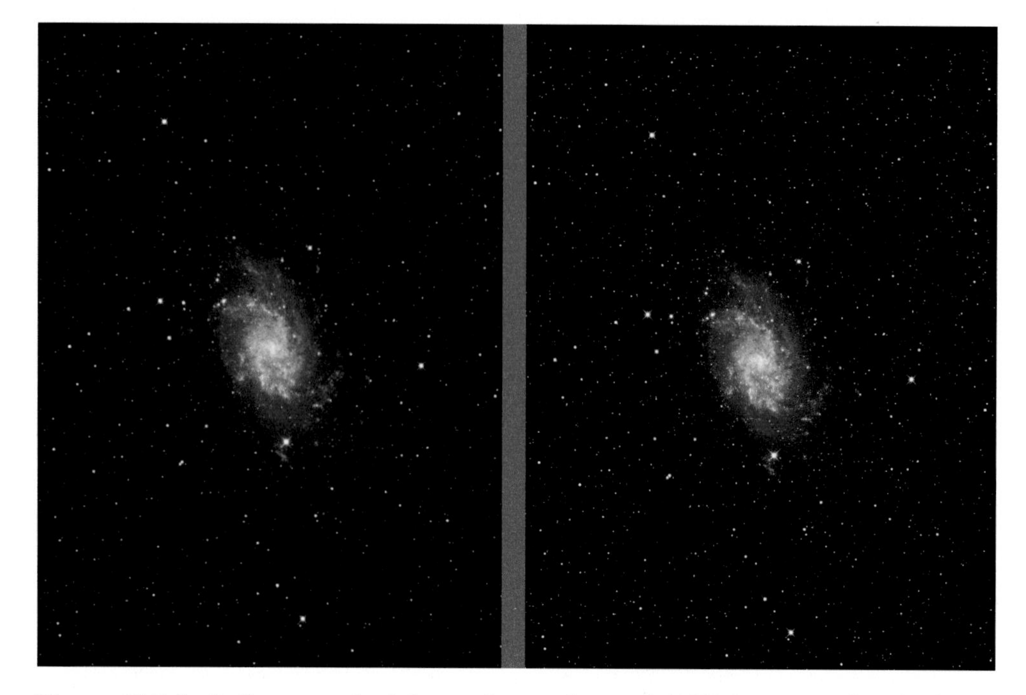

Figure 21.7 Left: the smoothed, but colour enhanced, RGB image; right: the luminance image

duplicate layer made (Layers > Duplicate Layer) with the opacity left at 100 per cent so that the copy is seen. This is simply so that the effect of the tool can be easily seen. The tool is entered by using: Image > Adjustments > Match Color. In the box that opens up, the 'Color Intensity' slider is moved to the right and adjusted (probably in the range 130–150) to get the required effect. By switching the adjusted layer on and off by clicking in its 'eye', one can easily compare the before and after images. With the enhanced image visible at 100 per cent, flattening the image (Layers > Flatten Image) will give the enhanced image.

It may well be that one only wishes to enhance the colour in a particular area of the image. This can be done by adding an adjustment mask to the enhanced image layer. The layer is highlighted by clicking in the Image box and masked using: Layer > Layer Mask > Hide All. The layer now appears black and only the original image is seen when both are selected. The 'Paintbrush Color' is now set to white and the area that is to be enhanced is painted over, revealing the enhanced colour in the painted area. (Black conceals and white reveals.) The transition from the original to the enhanced region is likely to be too obvious and this transition can be 'smoothed' by applying a Gaussian Blur of perhaps 50 pixels to the mask: Filter > Blur > Gaussian Blur. The two layers are then flattened as above to give the final image.

The final procedure is to give colour to the luminance image. One can either open the L and RGB images, copy and paste the luminance image over the colour one and

Figure 21.8 The LRGB colour image of M33

use the 'Luminance mode' to combine them or paste the colour image over the luminance image and use the 'Colour mode'. Most astroimagers tend to use the former, but I cannot detect any obvious difference between them. The Opacity slider can be adjusted to give the best result. Again, it may be necessary to align the two images as described above when combining the RGB images.

That's it! If the image is noisy, the 'Despeckle' filter in *Photoshop* may help somewhat or one can use a noise reduction program such as *Picture Cooler*. The final result of this imaging exercise is shown in Figure 21.8.

22
Narrowband Colour Imaging

In Chapter 20, the use of a narrowband H-alpha filter to produce monochrome images of the America, Pelican and Rosette Nebulae were described. By the use of further narrowband filters, colour images can be derived either in what one could call true colour – where one is trying to create an image that one could, in principle, observe or in 'false colour' where the objective is to create a beautiful image but one that bears no relation to what one could possibly see. This latter approach has become very popular.

Combining images taken through narrowband filters enables high quality results to be obtained even under light polluted skies or when the Moon is brightening the sky. Often three narrowband filters are used: (1) for the two S II lines in the deep red at ~7620 angstroms; (2) for the H-alpha line at 6563 angstroms in the red; and (3) for the two O III lines at ~5000 angstroms in the green. The three images (which will of course be monochrome if a monochrome CCD is used) are then allocated individual colours. In the so called 'Hubble Palette', the deep red Sulphur line is assigned to red, the red H-alpha line assigned to green and the green oxygen lines assigned to blue. The false colour images that result can look very beautiful, as in the famous 'Pillars of Creation' image taken by the Hubble Space Telescope. A further colour palette giving more natural colours assigns the H-alpha line to red, the O III lines to green and the S II lines to blue. Another palette assigns H-alpha to red and O III to green but uses a mix of the two to provide the blue layer. This reduces the overall imaging time as the S II lines are weak and so long exposures are needed. To equalise the brightness in the three images, the S II layer will tend to have to be 'stretched' more than the other two and this tends to increase the star sizes in the image giving a coloured 'halo' about the resulting stars.

Being an astrophysicist, I have never been too happy with the Hubble Palette, but I do have to say that it does sometimes bring out details that are harder to spot in RGB images. Owing to the fact that most of the light pollution lies towards the red end of the spectrum, I decided to use a variant of narrowband imaging that can be used to give a 'true' colour image provided that there was not too much moonlight. This is to use the H-alpha band to provide the red within the image, the O III lines for the green and to use a dichroic blue filter for the blue. The latter has a very sharp cut-off at the long wavelength end of its passband. My experiments were done when all the street lighting in my town was based on sodium lamps, but now many of these have been converted to LED lights, which have a far broader spectrum and so the technique may become less useful with time. However, as shown in Chapter 12, light pollution can usually be removed from the blue image – but it will still hide faint nebulosity.

Imaging the Dumbbell Nebula, M27, Using Narrowband Filters

I chose to image M27, the Dumbbell Nebula in Vulpecula, using a Vixen VC200L astrograph as it provided a suitable field of view when used with the SBIG ST8300 CCD camera as seen in the plot shown in Figure 22.1 produced by the *Astronomy Tools* Field of View Calculator: http://astronomy.tools/calculators/field_of_view.

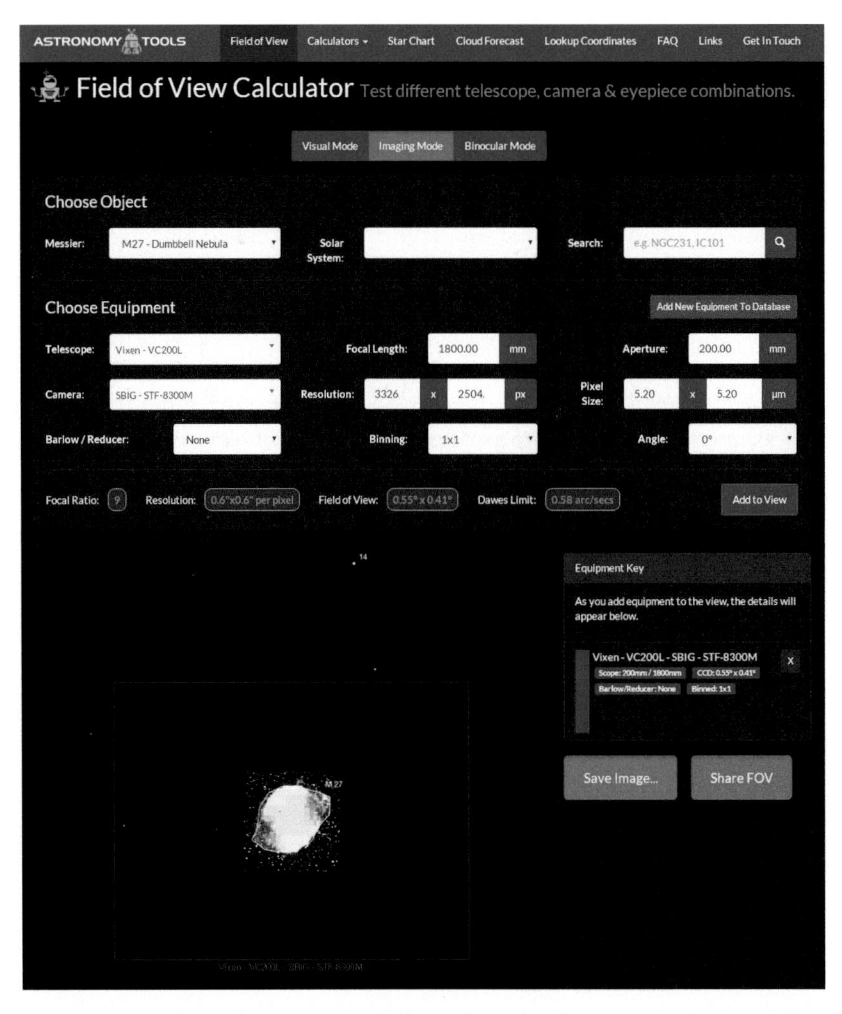

Figure 22.1 The *Astronomy Tools* Field of View Calculator shows the field that would be imaged with a Vixen VC200L telescope and SBIG ST8300 CCD camera

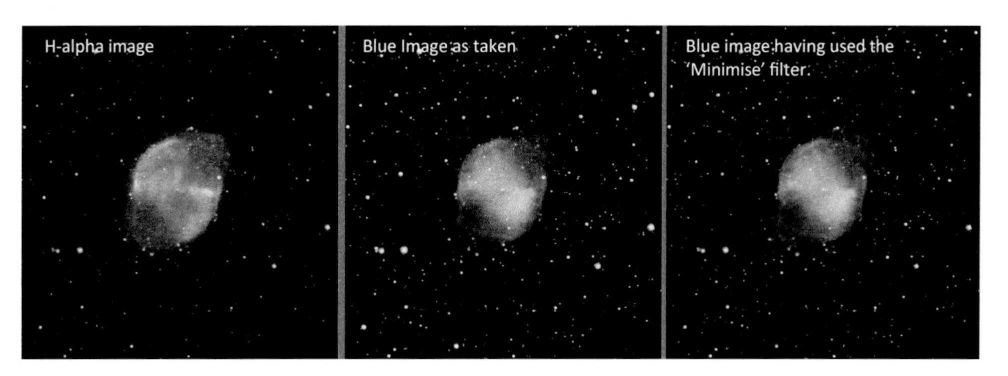

Figure 22.2 Using the 'Minimise' filter to reduce the size of stellar images

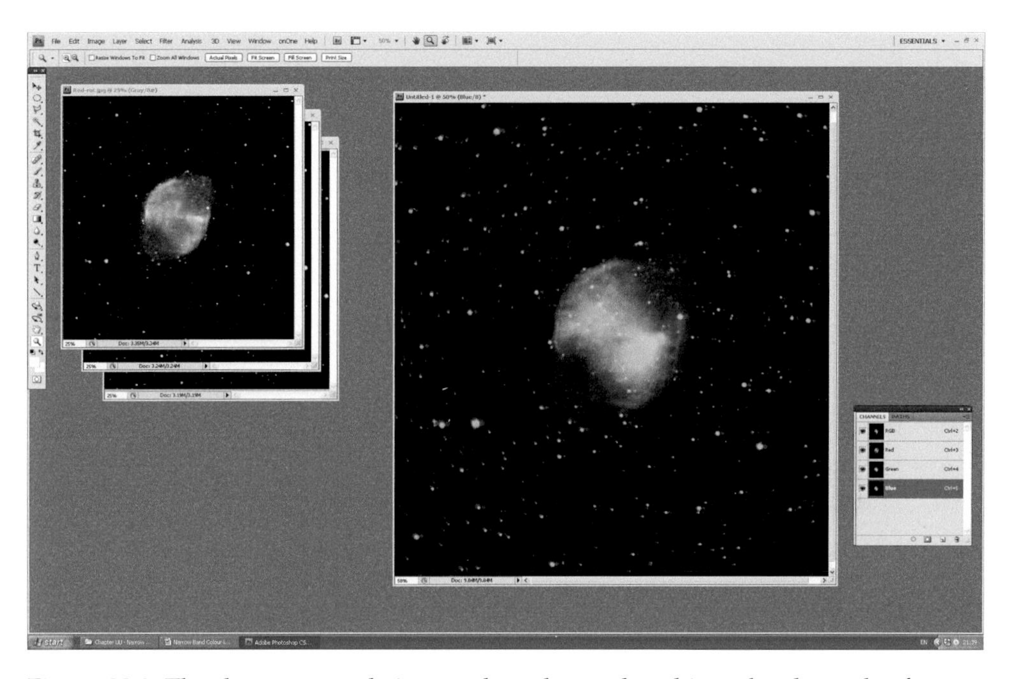

Figure 22.3 The three greyscale images have been placed into the channels of a new image to provide the colour image of the nebula. The colours are misaligned

M27 is quite bright, so the individual exposures, made up of 1 minute sub-frames, did not need to be too long and 30 minute total exposures in the three bands were all that was needed. A further 20, 1 minute dark frames were taken with the same sensor temperature of −20° Celsius. Each set of sub-frames was aligned and stacked in *Deep Sky Stacker* and exported as a TIFF file. Processing was identical with that employed in the LRGB example in the previous chapter, with the TIFF files stretched by several applications of a gentle curve and the background noise removed by bringing up the black point within the Levels command.

Looking at the individual RGB images, it was seen, as shown in Figure 22.2 left (H-alpha) and middle (blue), that the image of the central white dwarf star in the H-alpha image was significantly tighter – it would be expected to be less bright as white dwarf stars are very hot and so would emit far more blue light than red (especially as the H-alpha line is in the deep red). To enhance the final image somewhat, the 'Minimise' filter was used to tighten the stars in the blue and green images, with the result of having applied it to the blue image shown in Figure 22.2 right.

Figure 22.3 shows the result of initially placing the three greyscale images into the appropriate channels of a 'new' similarly sized RGB image – as described in the previous chapter. As is often the case, the alignments were not perfect and the blue

Figure 22.4 The final image of the Dumbbell Nebula taken by combining images taken with H-alpha, O III and blue filters

and red grayscale images had to be rotated and translated slightly to achieve perfect alignment with the green image.

The individual levels in the resulting image were adjusted slightly to improve the colour balance and the brightest stars, which looked a little 'hairy', were smoothed with a little Gaussian Blur to give the final image shown in Figure 22.4. I have to say that this is not significantly different to the image taken with a DSLR using a light pollution filter as described in Chapter 12, but one thing in the image that pleased me was the colour seen at the heart of the nebula. Once using a 16 inch Meade telescope at the Isle of Man Observatory under very dark skies and following very heavy rain which had left the sky transparent, I and others were able to see that the centre of the nebula was a beautiful iridescent green, not dissimilar to that seen in this image.

The Rosette Nebula Imaged Using the Hubble Palette

My colleague Christopher Heapy, who lives near me in a light polluted location and whose H-alpha image (Figure 22.5) of the Rosette Nebula in Monoceros was included in Chapter 20, used the Hubble Palette to produce this wonderful image of the nebula.

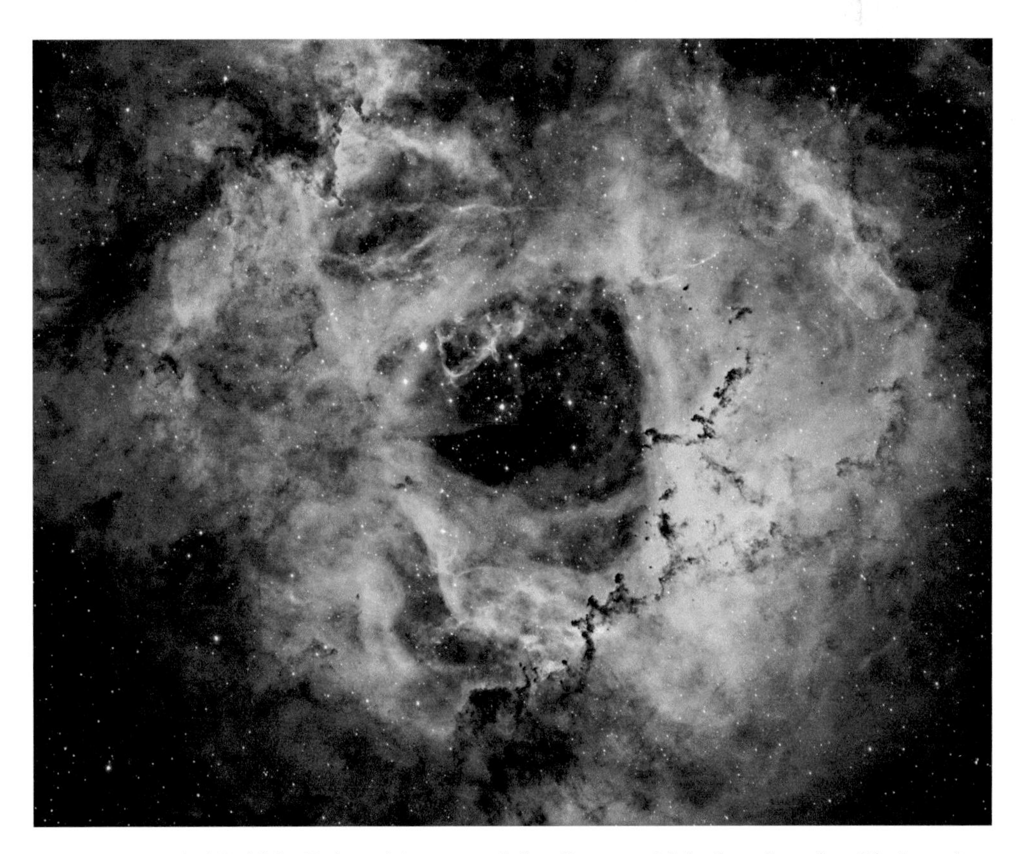

Figure 22.5 A 'Hubble Palette' image of the Rosette Nebula taken by Christopher Heapy

Figure 22.6 The red coloured H-alpha image showing Barnard's Loop and, near the top, the Angel Fish Nebula

Combining RGB and H-alpha Images

Given a 2-inch H-alpha filter and a modified Canon camera or, ideally, a monochrome CCD camera, it is possible to add an H-alpha layer to a RGB image taken with a DSLR. As an example, I chose to image the Orion constellation which includes several regions of H-alpha emission. To capture this, I used my SBIG monochrome CCD camera with a Nikon 35 mm, f/1.8, lens and Baader 2-inch H-alpha filter. The RGB image was captured with a Nikon full frame D610 camera and a 55 mm, f/2.8 Micro-Nikkor lens. Both systems give almost identical fields of view and the lenses were used (and are sharp) at full aperture.

Eighteen 1-minute, exposures (in 16-bit TIFF mode) were taken using the SBIG camera and stacked using *Deep Sky Stacker* using the median stacking mode to remove some aircraft trails. Forty 30-second, raw exposures were taken using the Nikon camera and similarly stacked.

The resulting H-alpha image was scaled to the size of the RGB image and its noise reduced using the *Picture Cooler* noise-reduction program (denoiser.shorturl.com). The stars were removed using the 'Dust and Scratches' filter with a pixel size of 10 pixels with any remaining stars cloned out. A colour layer of the same size was made using the colour (R = 200, G = 40, B = 40) and used to add colour to the H-alpha image to give the result shown in Figure 22.6.

This was then added into the RGB image using the 'Screen' blending mode with the use of the 'Transform' tool to rotate and move it to precisely overlie the RGB image giving the final result shown in Figure 22.7.

Figure 22.7 The final result of adding the H-alpha image into the RGB image using the 'Screen' blending mode

Appendix A
Telescopes for Imaging

Telescope Basics

Focal Ratio

A telescope tube assembly will have an objective of a given diameter, D, and have a focal length, F. The ratio of the two F/D is called the focal ratio. Typical focal ratios range from 4 to 15. It is easier to design an optical system with a larger focal ratio, so telescopes whose focal ratios are towards the lower end may need more complex – and hence expensive – optical designs or, as in the case of a short focal ratio Newtonian telescope, additional corrector lenses to make them suitable for astroimaging.

Telescope Magnification

The magnification is given by F/f_e where f_e is the focal length of the eyepiece. For example, when a 13 mm eyepiece is used with my 820 mm refractor, the magnification is 820/13 = ×63. It is often thought that the prime purpose of a telescope is to give very high magnifications, but the highest useful magnification is limited theoretically by the objective diameter (which defines its resolution) and practically by the 'seeing' at the time. This relates to the turbulence in the atmosphere above the telescope and usually limits the effective resolution to 2–3 arc seconds (an arc second is 1/3600th of a degree).

Field of View

When used with a sensor of a given size, the field of view in radians is given by its horizontal and vertical dimensions divided by the focal length of the telescope. These numbers are multiplied by 57.3 to give the field in degrees. So for my 18 × 13 mm Kodak sensor and 890 mm focal length refractor, the field of view is 1.15 × 0.84 degrees. An excellent field of view calculator can be found at: http://astronomy.tools/calculators/field_of_view. This allows a very wide range of cameras and

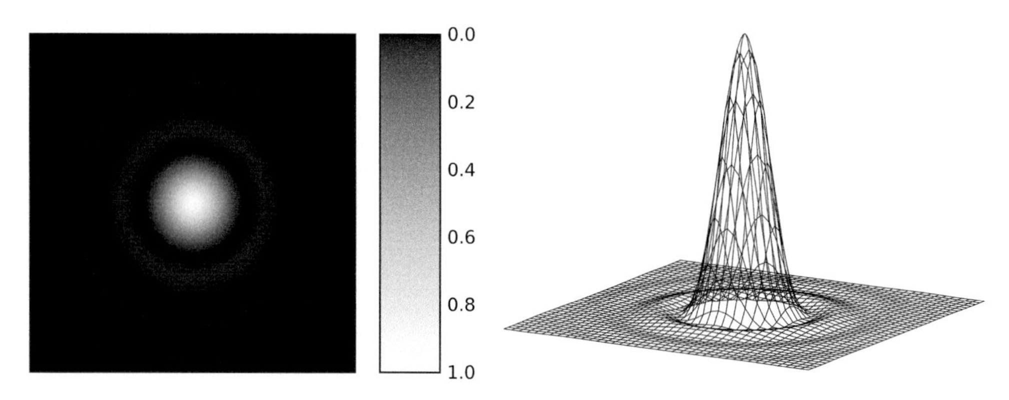

Figure A.1 The Airy pattern shown (left) as a greyscale pattern and (right) as a 3D plot. (Wikimedia Commons)

telescopes to be chosen and then derives the field of view and produces a chart showing how this might cover, for example, a chosen Messier object.

The Resolution of a Telescope

The detail in an image viewed by a telescope is theoretically limited by its resolution, which increases with telescope aperture. This is usually limited by the atmosphere though the nominal resolution can be approached under excellent seeing conditions and with the use of 'lucky imaging' – selecting and stacking the sharpest frames from a video sequence. The resolution of a given aperture scope can be measured experimentally by, for example, observing when the two stars of a close double can just be split. This approach gave rise to the empirical 'Dawes Limit' as proposed by W.R. Dawes and gives the resolution, in arc seconds, as $R = 4.56/D$, where D is in inches, or $R = 116/D$, where D is in mm. For a 100 mm aperture scope this gives 1.16 arc seconds.

The image of a star under perfect observing conditions and when using a telescope such as a refractor which has an unobstructed aperture is in the form of a central disc – called the Airy disc – which contains 84 per cent of the light, surrounded by a number of concentric rings of decreasing intensity. The whole is called the Airy pattern (Figure A.1). This is the result of the diffraction of light as it passes through the telescope aperture.

The Rayleigh Criterion states that a telescope can resolve two stars when the peak of one star's diffraction pattern falls into the first minimum of the other (Figure A.2). This gives a somewhat lower resolution limit than that defined by Dawes, as in the Raleigh Criterion there is a drop of ~26 per cent in brightness between the two peaks, whereas in the case of the Dawes Limit the drop is only 5 per cent. The angular separation between the centre of the Airy disc and the first minimum of the Airy pattern is given, in radians (1 radian = 57.3 degrees), by $1.22\,\lambda/D$, where λ is the wavelength of the light (both λ and D must be in the same units). Using the wavelength of green light of 5.5×10^{-7} m, this gives a resolution, in arc seconds, of $138/D$, where D is measured in mm. Thus, with a lens of aperture 100 mm, one gets a theoretical resolution of ~1.4 arc seconds.

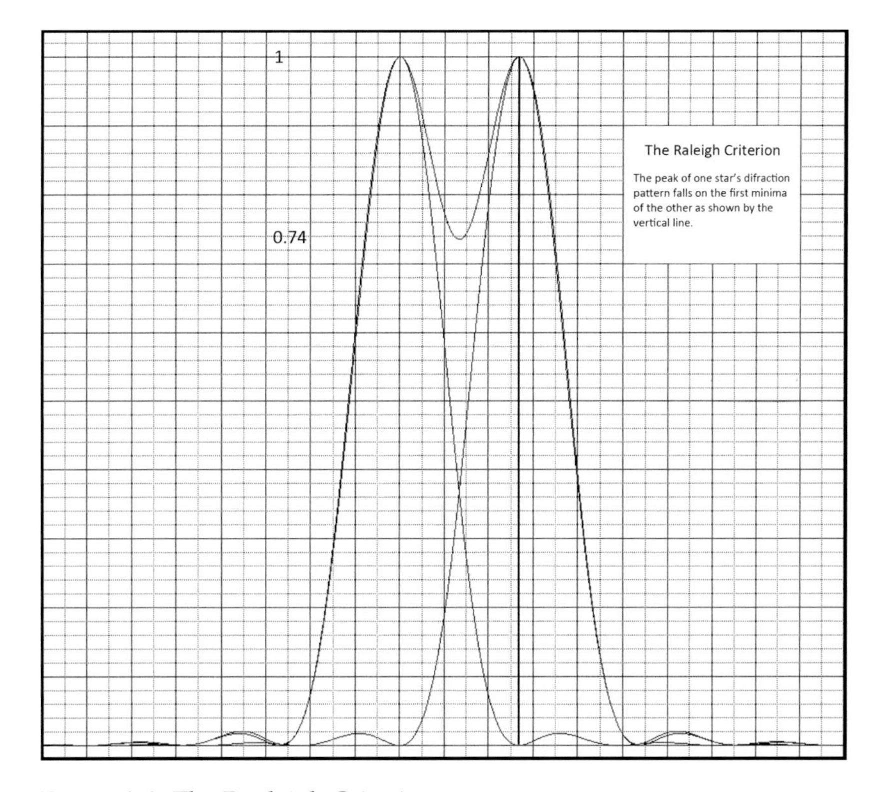

Figure A.2 The Rayleigh Criterion

Table A.1 *The theoretical resolution in green light for a number of telescope apertures.*

Aperture	Resolution
102 mm	1.35 arc seconds
150 mm	0.92 arc seconds
200 mm	0.69 arc seconds
300 mm	0.46 arc seconds

The Contrast in a Telescope Image

Image contrast is perhaps one of the key properties of a telescope and one that is particularly important when imaging the Moon and planets. It is a subject that is not too well understood, with erroneous statements often appearing in the astronomical press. The following two sections will, I hope, allow you to understand the various elements that come into play to determine what is termed the contrast of an astronomical image. The approach that I believe gives the best understanding of the subject splits the discussion into two parts: first that of the overall contrast of the image and secondly what I term the 'micro-contrast' of an image. I am not aware of

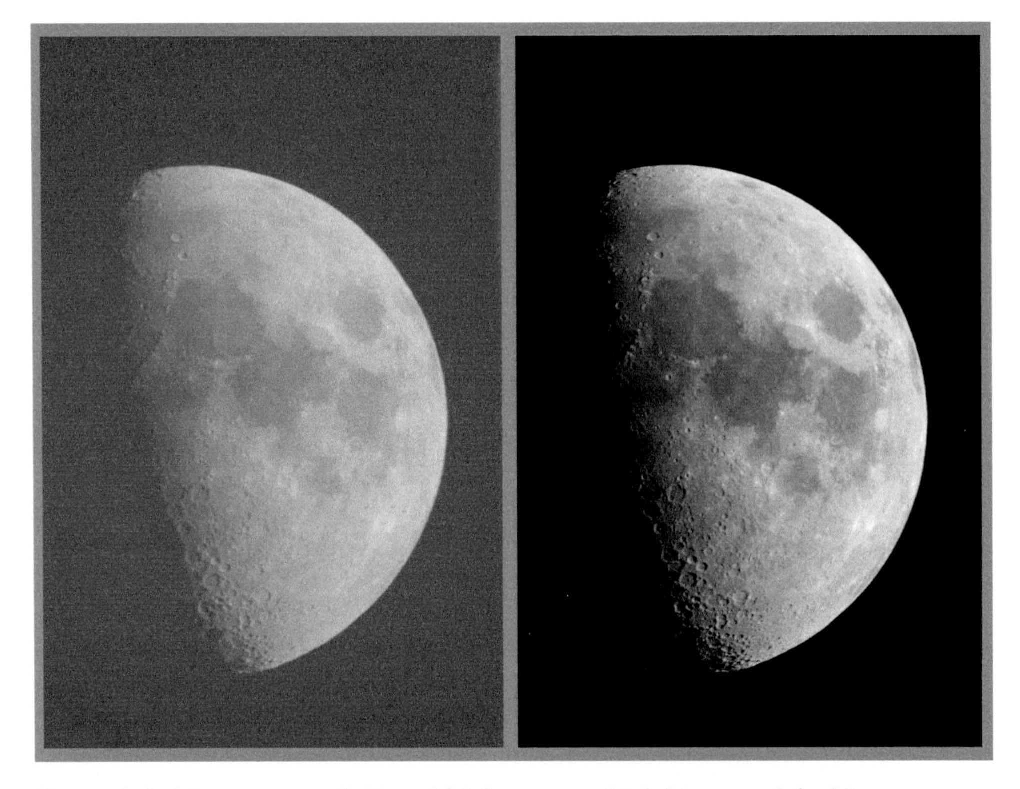

Figure A.3 A low contrast (left) and high contrast (right) image of the Moon

any other author taking this approach, but I honestly believe that this is by far the best way of considering this very important aspect of telescope design.

The Overall Contrast of an Image

So what is meant by the overall contrast of an image? In an ideal world, the light that is recorded by a CCD camera of a particular feature in an image would only have come from the light emitted or reflected by that feature alone. This will rarely be the case. The lunar image on the left of Figure A.3 is obviously of low contrast – the blacks are grey, not black. In this case, the reason was the fact that it was taken in twilight and so sky light was falling uniformly across the image. (By taking a sky image with the same exposure just to the left of the Moon and subtracting this from the lunar image, I removed much of this unwanted light and produced the image on the right, which has much higher contrast – quite a good tip!) However, after dark, there may well be light pollution spreading light into the image and, even with no light pollution, there will still be some 'air glow' reducing the contrast a little.

The overall image contrast (so that blacks are not as black as they should be) can also be reduced by factors in the design of the telescope being used. One often reads that reflecting telescopes give lower contrast images than refractors, but one major reason for this statement no longer holds quite so true. A telescope mirror having a simple aluminium coating will reflect ~86 per cent of the light falling upon it. As reflecting telescopes will have two mirrors, only ~75 per cent of the light entering the telescope will reach the eye or camera. This will reduce the effective brightness of the image somewhat but not, in itself, reduce its contrast. But what of the remaining 25 per cent? Some of this light will be absorbed within the mirror coating, but a significant portion will be scattered and can fall anywhere within the image, so reducing its overall contrast. (If you shine a red laser beam at a mirror surface so that the reflected beam is away from your eyes, you will easily see where the beam meets the surface, thus showing that light is being scattered.) A major reason why refractors give images with higher overall contrast than reflectors is that objective lenses may only scatter ~2 per cent of the light passing through them.

This is why I believe that the high reflectivity coatings that are now applied to many astronomical mirror surfaces are so important. With ~95 per cent reflectivity, not only will they give somewhat brighter images but they will also greatly reduce the amount of scattered light, so improving the overall contrast. A high reflectivity coating is well worth having even if at an additional cost: not only will the telescope perform better, but a second advantage is that the mirror surface allows far less moisture to penetrate and is likely to last perhaps 25 years before it needs to be re-coated. I have a 14-year-old Newtonian whose mirror was one of the first to be given a high reflectivity coating and it still looks like new. The 'HiLux' coating given to this mirror was applied by Orion Optics (UK) and they can recoat any mirror with this multilayer coating if desired.

A second cause for reduction in the overall contrast is when light scatters off the interior of the optical tube assembly. This is also why refractors can provide such high contrast images, as a series of knife edge baffles reducing in size can be located within the optical tube to trap any scattered light. Matt black flock coatings may also be used in both types of telescope to reduce scattering, and high specification reflecting telescopes may also be equipped with a series of baffles mounted within the tube. If a reflecting telescope has a glass correcting element mounted at the front of the tube assembly, it is a good idea to use a dew shield (which extends the tube assembly outwards), not just to reduce the building up of dew on its surface but to prevent extraneous light falling upon it. The best of these are also equipped with internal baffles. Even with a Newtonian, an outward extension to the tube will help when extraneous light is a nuisance.

The design of the telescope will affect the overall contrast as well. It is impossible to beat a well-designed refractor, but Newtonian telescopes, where one observes across the tube assembly to the far wall are almost as good. This also applies to the more complex Schmidt–Newtonians and Maksutov–Newtonians. My 150 mm Maksutov–Newtonian (Figure A.4) has a set of baffles immediately across from the focuser to prevent any light scattered off the tube walls entering the field of view. Few standard Newtonians seem to be so equipped, and the application of some flocking opposite the focuser could well make a useful improvement.

Figure A.4 A Maksutov–Newtonian showing baffles opposite the focuser

The telescope designs that have the greatest problem with overall contrast are those where the light path exits through the primary mirror, as one is then looking up towards the sky. Such telescopes incorporate an internal baffle tube so that the incident light into the telescope is hidden by the secondary mirror and its support. This does involve some design compromises as extending the baffling to increase the overall contrast may well restrict the light falling on the outer parts of the image – called vignetting. Even so, extraneous light can still enter the baffle tube and be scattered into the image. Again a dew shield will greatly help. Increasing the size of the circular secondary mirror support will also help, but this then impacts on the second cause of reduced contrast within an image which I term 'micro-contrast', as discussed below.

The Micro-contrast of an Image

As described above, the overall contrast will be reduced by *scattered* light either from parts of the optical tube assembly, from the surface of a mirror or (to a far lesser extent) within the objective lens. The micro-contrast, on the other hand, is determined by the effects of light that is *diffracted* by parts of the optical tube assembly. In this case, light is only moved from its rightful place by angular distances measured in arc seconds or arc minutes and is thus particularly important in the case of observing planetary discs where the angular scale of the observed object is similar in scale, and the features on the surface may have low contrast as well.

Table A.2 *The energy within the central Airy disc as a function of the relative size of the central obstruction.*

Percentage central obstruction	0	10	20	30	40	50
Energy in central disc (%)	84	82	76	68	58	48

The Effects of a Central Obstruction on the Airy Pattern

The first effect of diffraction is that caused by the fact that the telescope will have an aperture of a given size. The result is that a point source of light such as a star gives rise to a disc of light (whose size is determined by the aperture of the telescope) surrounded by concentric rings forming the Airy pattern, as has been described above. If the aperture is unobstructed, as with a refractor, 84 per cent of the light falls within the central disc and thus 16 per cent lies in the rings – with the majority within the first ring, whose diameter is about twice that of the central disc. This changes when the aperture is obstructed by a central obstruction such as the secondary mirror in a reflecting telescope. As the obstruction increases as a percentage of the aperture, more light is transferred from the central disc into the rings. Table A.2 shows how much light remains in the central disc as the percentage of the aperture obstructed by the secondary mirror increases. At the same time, the angular diameter of the central disc actually reduces slightly, so possibly helping to resolve double star systems. The actual energy in the central disc relative to the maximum possible (84 per cent) is called the Strehl ratio and is discussed more fully later in this appendix.

It is generally reckoned that an obstruction of up to 15 per cent has virtually no observable effect. The size of the secondary in a Newtonian or Maksutov–Newtonian (M–N) is reduced as the focal ratio is increased and there are scopes of this type specially optimised for planetary viewing. An f/9 Newtonian and some M–N scopes have secondary mirrors giving an obstruction close to 15 per cent. The very small secondary in an M–N telescope optimised for planetary imaging is shown in Figure A.4 above.

A way of improving the effect of the secondary mirror is to reduce its size so that only the very central part of the field of view is fully illuminated. That is, if you place your eye (without an eyepiece) at the centre of the focuser you will see all of the mirror, but, as you move away from this central position the area of the mirror that you see will be reduced. Thus less of the mirror will be illuminating these points in the field of view, so causing vignetting away from the centre of the field. This is, of course, no problem when viewing the planets, provided that they are kept in the centre of the field of view. It is possible to have two changeable secondary mirrors, a small one for planetary observing and a larger one for deep sky use to more fully illuminate the whole field of view.

A second diffraction effect is found when the secondary mirror has to be supported by a spider as in Newtonian telescopes. This produces a thin cross which is

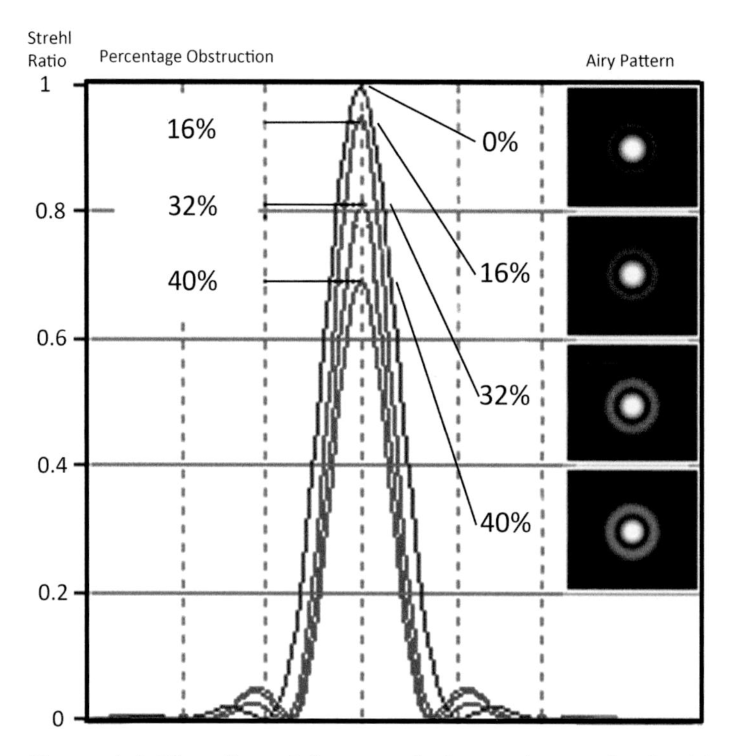

Figure A.5 The effect of the central obstruction on the Strehl ratio and Airy pattern

centred on stellar objects. In fact, the result is quite pretty and it has been known for astroimagers using refractors to stretch two strings across the objective to give the same effect as a spider! It is not obvious on extended objects such as a planetary disc, but the effect is still present and will reduce the micro-contrast of the image. By using a curved spider the spikes disappear as the light is spread around more uniformly, but light will still be scattered away from its rightful place.

Finally, it is often said that Schmidt–Cassegrains (S–Cs) are not good for planetary observing as they have a large central obstruction. Figure A.6 shows how the effective resolution of a 200 mm telescope falls off as the central obstruction increases in size. Unless the seeing is near perfect, the effective size of the central disc is given by the diameter of the first ring in the diffraction pattern. This is about twice the diameter of the central disc of an unobstructed aperture, so the effect is to halve the nominal resolution. True, but your S–C is likely to be at least 8 inches in diameter and so will still have the same resolution as a 4 inch refractor – and no one says that *they* are no good for planetary observing. But the image will be far brighter and this may allow greater magnification to be used. When the seeing is superb, the fact that the diameter of the central disc reduces slightly as the central obstruction becomes larger may actually increase the resolution of a telescope and the highest resolution Earth-based images of the planets have been made with large aperture Schmidt–Cassegrains.

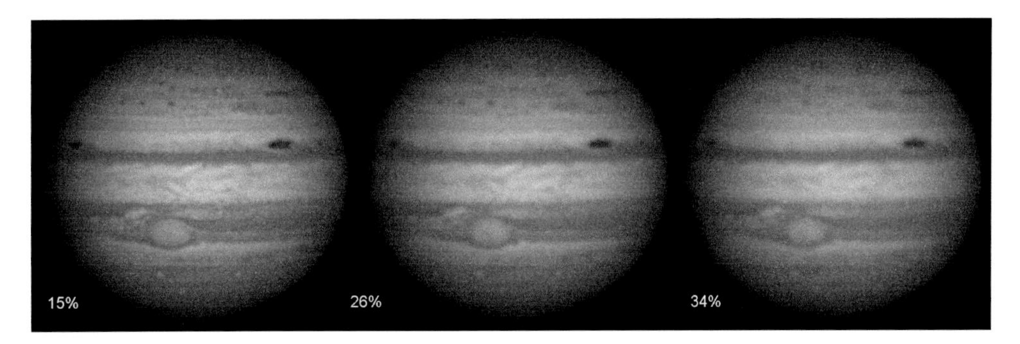

Figure A.6 The effect of a secondary obstruction on the resolution of a telescope

Image Quality

One factor in determining the image quality of a telescope is the precision with which the mirrors or lenses have been made. Refractors where the false colour (chromatic aberration) has been eliminated are often said to give exquisite 'pin sharp' images, whereas those produced by reflecting telescopes, particularly Schmidt–Cassegrains, are often said to be somewhat 'mushy'. There are several reasons why this might be so.

1) A refractor is usually perfectly collimated – that is, optically aligned – and tends never to need collimation, whereas reflectors to tend to lose their collimation and need to be regularly collimated to give of their best.
2) The central obstruction of a reflecting telescope will reduce the image quality somewhat compared to an identical aperture refractor.
3) The overall contrast of refractors will tend to be higher.
4) A refractor is likely to have an optical system of higher accuracy and so closer to optical perfection. The reason is very simple. It is actually four times easier to make a lens with a given optical precision than a mirror. Suppose a mirror has a 'perfect' surface except for a small part which is 1/8th wavelength below its surroundings. As the light is reflected, the error in the light path length will be double this and so be 1/4 wavelength. The error in the path length for any deviation from a perfect surface is doubled. Let's now consider the surface of a lens that has a depression of 1/8th wavelength. Had it been filled with glass, the difference in optical path length would have been increased by the fact that the glass has a refractive index of ~1.5 – but this is only by ~50 per cent more than the path though air. So the actual path length error will only be 1/16th wavelength – one quarter that of the equivalent error in a mirror!
5) The expensive refractors that are prized for their exquisite images tend to have more time expended on them by a master optician than is generally the case with mirrors and so will tend to be closer to optical perfection.

This is not to say that reflecting telescopes cannot be superb and one can purchase mirrors, such as those made by the Zambuto Optical Company, which are very close to optical perfection and give images comparable to the very best apochromat refractors. Happily the vast majority of telescope optics are now excellent and, at a Scottish star party in January 2013, the image of the central region of the Orion nebula as seen through a Celestron 11 inch Schmidt–Cassegrain was the best I have ever seen – the stars making up the trapezium showing as perfect pinpricks of light. A well collimated Newtonian or Schmidt–Cassegrain whose mirror has high reflectivity coatings can be a superb telescope.

How Is the Quality of a Mirror or Lens Assessed?

There are a number of ways by which a manufacturer of mirrors or lenses may specify the quality of the optics.

Strehl Ratio

This is, perhaps, the most obvious and best defined way of defining optical quality and one that is now coming into more common use. As described above, a refractor, as it has an unobstructed aperture, would ideally place 84 per cent of the light from a star (when observed under perfect seeing) into the central disc of the Airy pattern. If the objective were perfect, it would then be said to have a Strehl ratio of 1. No objective is perfect, although some of the very best reach Strehl ratios well above 0.95 under test conditions. Some manufacturers may be willing to quote minimum Strehl ratios for their objectives – a value such as 0.95 or 0.96 – but it is interesting to note that the very 'high end' manufacturers, whose lenses often equal or exceed a 0.98 Strehl ratio, may not do so for fear of competition among their purchasers. The important point to note is this: it is virtually always that the atmospheric seeing that will limit the image quality of a good telescope and a Strehl ration of 0.95 or above can essentially be regarded as perfect.

When a telescope has a central obstruction, the light within the central disc is reduced and so the Strehl ratio is bound to be less, and, if the actual Strehl ratio of the telescope were given (likely to be nearer 0.8 rather than 0.9), its value would not be representative of the inherent quality of the mirror. In this case, the Strehl ratio quoted is for the mirror assuming that there is no central obstruction. Values well above 0.9 may be then be measured and quoted. Figure A.5 also shows how the Strehl ratio reduces as the size of the central obstruction increases.

Root Mean Squared (RMS) Surface Accuracy

Mirrors can now be tested using interferometric methods and these yield another way of defining the surface accuracy. The RMS accuracy of a mirror provides a statistical measure of the departure of the surface from the ideal shape. The RMS

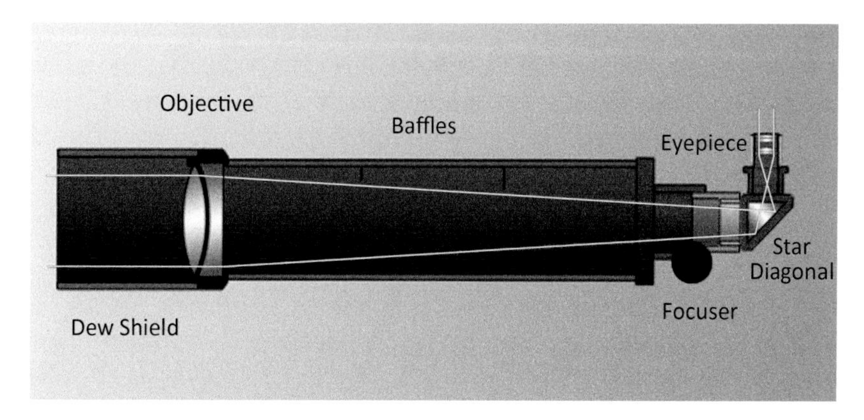

Figure A.7 The classic refractor design with doublet lens
(Image: Starizona)

value can be converted into the Strehl ratio of the mirror. The approximate formula is:

$$\text{Strehl Ratio} = 1 - (2\,\pi \times \text{RMS})^2$$

Putting in, say, 0.028 for the RMS value, multiplying by 2 π and squaring gives 0.032, resulting in a Strehl ratio of 0.97.

Mirror Smoothness

This is something that few manufacturers will specify, but, has a significant impact on the micro-contrast of an image. A pattern of ripples in the surface which may be less than 1/30th of a wavelength in amplitude (and so will not figure in the overall mirror specification) will, nevertheless, reduce the contrast of the features on a planetary disc. This is one area where hand finished mirrors made by the world's top opticians such as Carl Zambuto and Robert F. Royce outperform those from large scale manufacturers. The Zambuto Optical Company website gives considerable insight into this aspect of mirror performance.

Refractors

In 1733, an English barrister named Chester Moore Hall invented what is termed an achromatic doublet, which is composed of two lenses made from glasses having differing amounts of dispersion. Typically a convex lens made of crown glass is mated to a concave lens made of flint glass so that their chromatic aberration is cancelled out. The converging, positive power of the convex lens is not quite equalled by the diverging, negative power of the concave lens, so together they form a weak positive lens. This was for a generic pair of glasses. If the two types of glass are chosen with care, particularly if one has a very low dispersion or, even better, a lens made of calcium fluorite crystal, the correction can be amazingly good and the resulting doublet lenses are effectively colour free when in focus. Such doublet lenses are often called ED doublets, where ED stands for 'extra low dispersion'. Pure fluorite lenses are now rare, partly owing to cost but also because some glasses, sometimes called FD or SD glasses, produced in recent years have very similar properties to fluorite and can be used with an appropriate mating glass element to make lenses that are essentially as good.

Glass Types Used in Refractors

In the specifications of refracting telescopes mention will often be made of the glass types used to manufacture their lenses. There is usually a low dispersion (ED) glass used either as one element of a doublet or as the central element in a triplet. A key property of an ED glass is its 'Abbe Number', which determines how little dispersion is introduced as the lens refracts light. In summary, the types are as follows:

1) Fluorite crystal, CaF_2 with an Abbe Number of 94.99 – very expensive and hard to work with and now only found in relatively small diameter lenses.
2) FPL-53 'Super ED' or 'FD' glass made by Ohara in Japan with an Abbe Number of 94.94 – less expensive and easier to work with than fluorite crystal, with a very similar Abbe Number to fluorite. Chemically more stable than fluorite. Also referred to as synthetic fluorite or 'SF' glass. It is easier to design a colour free lens using FPL-53 than glass with a lower Abbe number, but it is somewhat harder to figure and does not maintain its shape as well with changes in temperature. Triplet lenses using fluorite or FPL-53 can give essentially perfect correction for chromatic aberration both in focus and outside of focus.
3) FPL-51 ED glass made by Ohara in Japan with an Abbe Number of 81.54 – less expensive and easier to work with than FPL-52 and chemically very stable. It maintains its shape better than FPL-53 with changes in temperature and this is an asset when apertures greater than about 115 mm are considered. The colour correction will not be quite as good as smaller lenses made using FPL-53 and may show a little false colour in out of focus images but essentially none in in-focus images – which is what really matters.
4) FCD1 ED glass made by Hoya in Japan with an Abbe Number of 81.6 – very similar to FPL-51.
5) H-FK61 glass made by CDGM in China with an Abbe Number of 81.6 – very similar to FPL-51.

It should be pointed out that the mating elements used to match the characteristics of the ED element are just as important in controlling chromatic aberration as is the accuracy with which the lenses are figured, coated and assembled. In general, FPL-53 doublets will perform better than FPL-51 doublets *of the same aperture and focal ratio* and triplets will control colour better than doublets *given the same aperture and focal ratio*. I have highlighted the fact that this is true only when a comparison is made of identical apertures and focal ratios. When surveying the wide range of refractors on offer, one sees that FPL-53 tends not to be used to improve the chromatic aberration

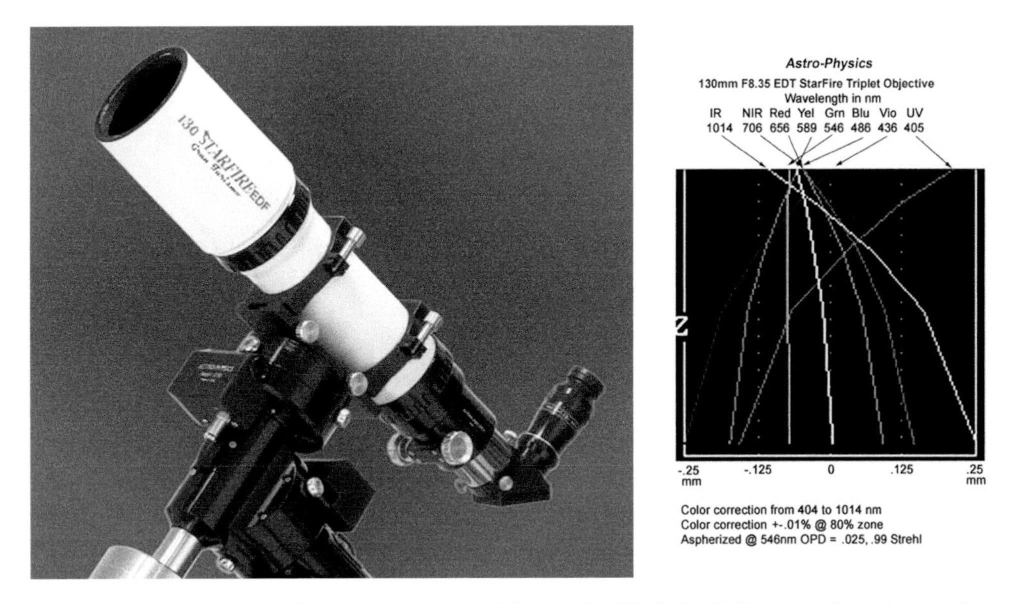

Figure A.8 An Astro-Physics 130 mm, f/5.3, STARFIRE EDF. The plot shows the superb colour correction and Strehl ratio of 0.99 (essentially perfect) of their earlier 130 mm, f8.35 refractor

present in an objective over that when FLP-51 is used but rather to reduce the focal ratio. So an FPL-51 doublet refractor might well have a focal ratio of f/6.8, while a telescope of the same aperture using an FPL-53 doublet would have a focal ratio of f/6 or f/6.25, with both exhibiting similar (low) levels of chromatic aberration. This is presumably as astroimagers prize low focal ratio telescopes! Triplet lenses using both FPL-51 and FPL-52 will tend to reduce chromatic aberration even further.

Achromats and Apochromats

Telescopes using generic glasses to make the standard achromatic doublet have been given a rather bad name of late. This is unfair. If a telescope having a focal ratio of f/12 to f/15 is made using a simple achromat, the chromatic aberration is minimal and, as the curvature of the lenses used to form the doublet will be less than for smaller focal ratios, the other lens aberrations, notably spherical aberration, will be less.

An achromat objective will bring two colours of the spectrum to the same focal point and should be corrected for both spherical aberration and coma at one focal point – usually within the green-yellow part of the spectrum, where the eye is most sensitive. Ernst Abbe of the Zeiss Company defined the specification of a more advanced lens which he termed an apochromat that will bring three colours to the same focal plane. Assuming that the design is done well, an apochromat can reduce the observed colour fringing to near zero. At the same time an apochromat should also be corrected for spherical aberration and coma at two widely separated visual wavelengths rather than one. This specification is very difficult to meet and may not even give the best optical performance. It is unlikely that any of the apochromat telescopes now available meet all parts of his definition.

The era of apochromat telescopes for use by amateur astronomers really began in 1981. An amateur telescope maker called Roland Christen had come across a batch of an abnormal dispersion flint glass that had been ordered by NASA but not used. Christen was able buy this glass and designed some triplet lenses using it for one element. His prototype five inch, f/12 refractor produced the best images of Jupiter of any telescope at the Riverside Telescope Making Conference and earned him the prize for the most innovative optical design. Not long after he gave up his well-paid job, set up a company named Astro-Physics, and went into the full time business of making telescopes. The refractors made by his company are some of the most prized telescopes now available and there is a long waiting list for people wanting to order one.

A second notable optical designer, Thomas Beck, of TMB Optics, also designed a series of very high quality refractors with their triplet objectives being made by the Russian optical company LZOS and mounted in tube assemblies made by APM in Telescopes Germany. Sadly, Beck died unexpectedly in late 2007 but APM continue to provide superb apochromat refractors using lenses made by LZOS.

In Japan, Vixen and Takahashi designed doublet objectives where one element was made from a pure fluorite crystal. The Takahashi FS series have become a legend but only smaller aperture telescopes are still available partly owing to cost and environmental reasons. Fluorite has the lowest dispersion of any lens type, so when mated with a suitable glass element, superb colour correction can be made with just two elements. One cause of reduced contrast in an image is due to scattering of light within the objective lens. Fluorite crystal scatters virtually no light, so these telescopes are believed to show the highest contrast of any telescope. A Takahashi FS102, f/8 telescope is one of my most prized possessions and its superb contrast enabled me to produce an award winning image of 'Earthshine', as seen in Figure 5.5.

There is one other optical design that is used in high quality refractors: that of using four elements in two doublets, one acting as the objective and the second, nearer to the eyepiece, acting to correct the aberrations inherent in the objective doublet. This is called a Petzval arrangement. Televue, perhaps better known for their eyepieces, make the NP101is and NP127is, whilst Takahashi produce the FSQ85-EDX and the FSQ106-EDX. These are designed to give very wide flat fields of view and, with large focusers, can be used with the largest CCD cameras now available, so are perfect for astroimaging. Not unsurprisingly, these four-element refractors are probably the most expensive per inch of aperture of any telescope that can be bought. Quite a number of manufacturers are now producing refractors with 4 or 5 elements to give the wide, flat fields, much prized by astrophotographers. Examples are the Teleskop Service TSAPO65Q imaging refractor which uses a FLP-52 based triplet objective and singlet field flattener within the tube assembly to provide a flat field that will cover a full frame sensor and the William Optics Star 71 that uses a FPL-53 triplet objective and two further singlet lenses giving a similar sized flat field.

As the effects of chromatic aberration are reduced to near zero, other aberrations play a more important role. One, in particular, is called spherochromatism – the variation of spherical aberration with wavelength. The optical designer has to balance all these out to try to produce an objective lens of high quality at an affordable price. One reason why apochromatic lenses tend to have three elements is that the curvature of the individual lenses is less and this makes spherical aberration and spherochromatism less of a problem as well as increasing the tolerances in its alignment, so making fabrication easier.

One American company of note is the Telescope Engineering Company, or TEC for short, whose first apochromatic refractors came to the market in 2002. Perhaps the most famous of their line is the APO140, F/7, ED oiled triplet. In this objective, the three elements, rather than being air spaced, are separated by thin films of oil. This has some real advantages over an air spaced triplet: the oil filling any minute imperfections in the internal surfaces which may not even need to be finely polished. (Applying water to a piece of roughened glass, perhaps found on a shoreline, will make it appear smooth!) The middle element is made from Ohara FLP-53 and, as the objective has only two air glass surfaces at the front and rear, there is less internal scattering, so giving high contrast images. A further advantage of an oiled spaced objective is that the central element in the lens can cool more quickly to the ambient temperature and follow any reduction in temperature during the night rather better than an air spaced triplet. This is a telescope that many amateurs aspire to!

The European company CFF Telescopes also produce a range of oil spaced triplet refractors with guaranteed Strehl ratios greater than 0.96, but often exceeding 0.98. One element is made aspherical to reduce aberrations that are harder to eliminate than when an air spaced triplet is used.

Oil Spaced or Air Spaced Triplets?

Astro-Physics have used a number of oil spaced objectives in the past and their current 140 mm f7.5 Starfire EDF incorporates one, whereas their 175 mm f/8 Starfire EDF uses an air spaced triplet. The advantages of an oil spaced triplet have been described in relation to the TEC 140 mm refractor, but what advantages might an air spaced triplet have? In an oil spaced triplet, the two radii of curvature of the central element are determined by the inner radii of curvature of the outer elements. By separating the three lenses, the designer can alter the radii of curvature of this central element and this gives them two more degrees of freedom with which to optimise the design, so perhaps being able to reduce the spherical and spherochromatic aberrations of the lens. The modern multi-coating that can be applied to the lens surfaces helps to minimise any internal reflections or scattering so that the difference in this respect compared to an oil spaced objective is now less than in the past. Their one problem is that it is more difficult for the central element to cool and thus more time will be needed for them to cool down to ambient outside temperatures. If the air temperature is dropping rapidly after nightfall, it can even be possible that an air spaced triplet will never come to thermal equilibrium.

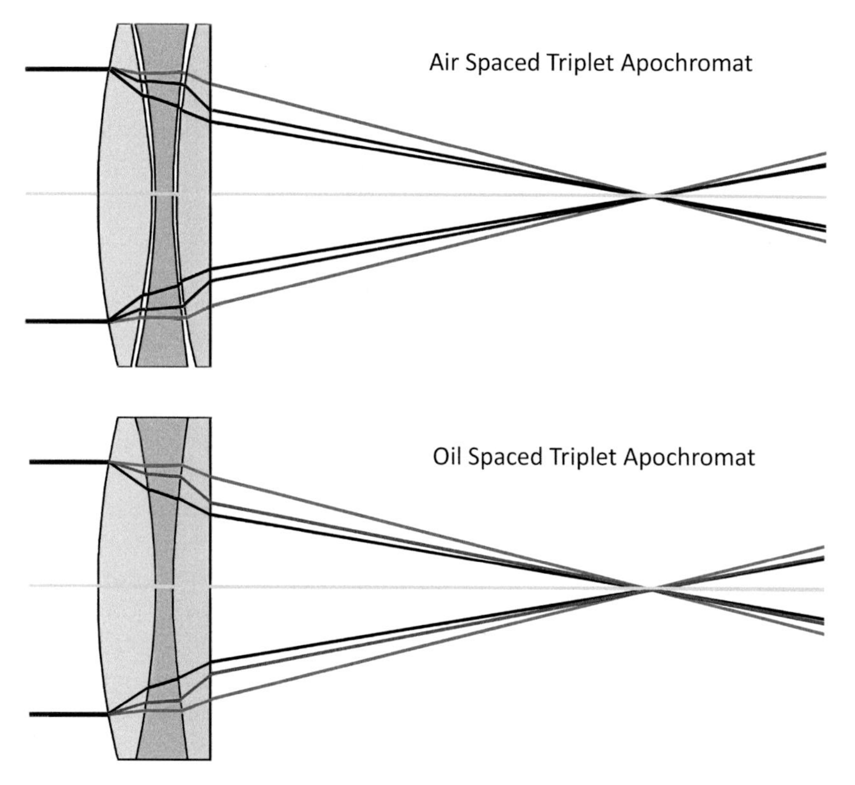

Figure A.9 Air and oil spaced apochromats (not to scale)

Focuser

The focuser barrel diameter is important when the telescope is to be used for wide field astroimaging: too narrow and it will vignette the outer parts of the field of view when large sensors are used. So, as well as the 'standard' barrel size of 2 inches, focusers having larger diameters such as 3.5 and even 4 inches may be provided as standard, or available as an option. For many years, the Crayford focuser has been supplied with high quality telescopes, but they can have a problem with supporting CCD cameras and filter wheels, particularly when imaging near the zenith. For this reason many refractors are now being equipped with helical rack and pinion focusers, but, as they require precision machining, they tend to be quite expensive. The Starlight Instruments Feather Touch rack and pinion focusers, as used on my CFF Telescopes 127 mm refractor, cost over £500!

Newtonians and Their Derivatives

Qualities and Drawbacks of a Newtonian Telescope

- They are free of chromatic aberration.

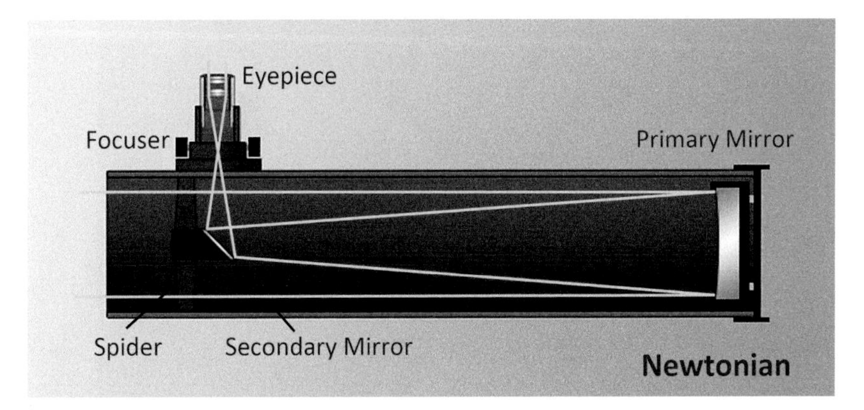

Figure A.10 The classic Newtonian design
(Image: Starizona)

- For a given aperture, they are usually less expensive that other telescope designs of comparable quality.
- Only one surface, that of the primary mirror, needs to be ground and figured into a complex shape.
- The image field suffers from coma away from the optical axis, causing stars to appear like little comets with a flare pointing towards the field centre. This gets greater the further away from the axis and is inversely proportional to the focal ratio so Newtonians with a low focal ratio, such as f/4 suffer the most. This is not the problem that it once was as 'coma correctors' are readily available to overcome this defect.

The Basic Design of a Newtonian

As shown in Figure A.10, the basic design of a Newtonian is very simple. At the base of the telescope tube a parabolic mirror is supported on a movable mount so that, using three adjustment screws, its optical axis can be aligned along the centre line of the telescope tube. Near the top of the tube a 'spider' supports a secondary mirror, which reflects the light sideways to the image plane (the focus of the mirror) just outside the telescope tube, where a focuser supports the eyepiece that is to be used. The secondary mirror is cutting across a converging cone of light and is thus a conic section which, in this case, is an ellipse and so the mirror is what is called an 'elliptical flat'. The designer of a Newtonian calculates the minor axis of the mirror and the major axis is simply the minor axis dimension multiplied by the square root of two, 1.414, so an elliptical flat having a minor axis of 50 mm would have a major axis of 71 mm. The secondary mirror forms what is called the secondary obstruction, whose diameter is that of its minor axis.

The size of the secondary mirror is a compromise. If too small, it will cause the outer parts of the field of view to be vignetted (that is darkened as compared to the

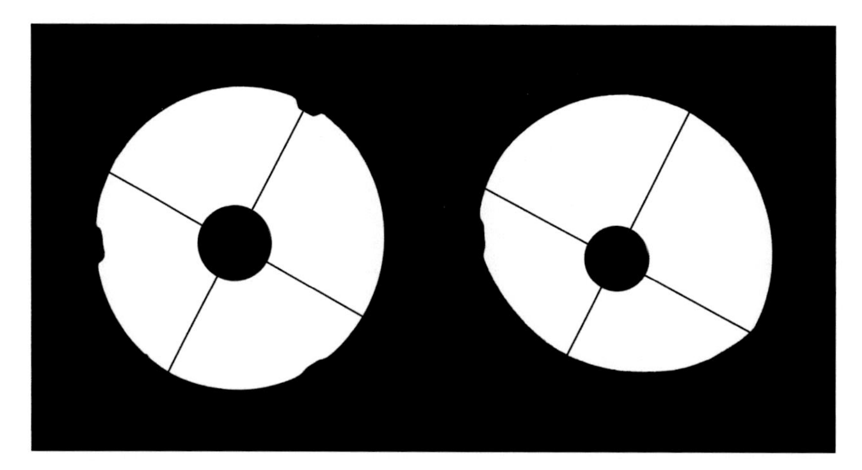

Figure A.11 The primary mirror of a 200 mm, f/6 Newtonian as photographed from the centre of the field of view (left) and from the edge of the 1.25 inch focuser (right)

centre), but the larger it is, the greater its diffraction effects, so reducing the micro-contrast of the image. The size of the secondary mirror will determine what percentage of the field of view will be 100 per cent illuminated. Figure A.11 shows the view of the primary mirror taken at the centre of the focal (or image) plane and as one moves towards to one side. From the centre of the field of view you must be able to see the whole of the mirror, but, away from the centre, one will tend to see less of the mirror, as shown in the figure. This means that less light will fall on that part of the field of view, so causing vignetting.

One might at first think that the obvious thing is to simply use a secondary that will fully illuminate the largest field of view that might be used, but the larger the size of the secondary, the more light is transferred from the central Airy disc into the surrounding rings, so reducing the micro-contrast in the image. In practice, it does not really matter if the illumination falls to 70 per cent of the maximum at the edge of the field of view when the telescope is used visually, though for wide field imaging with a DSLR, a greater percentage might be preferable. The optimum design of a Newtonian thus actually depends on the use to which it will be put.

A Newtonian Optimised for Planetary Imaging

To minimise the effects of the secondary mirror, the focal ratio should be greater than f/8 and ideally a low profile focuser should be used so that the distance from the image plane is minimised and thus the secondary mirror will be smaller. In fact, for this use, the secondary mirror can have a minimum size so that only the very central part of the image plane is fully illuminated – where, of course, a planet would be placed when imaging. A 150 or 200 mm Newtonian with a small secondary can be a superb instrument for planetary imaging with a webcam!

A Newtonian Optimised for Wide Field DSLR or CCD Imaging

To achieve a wide field, a short focal length telescope is wanted, so as low a focal ratio as possible will be best. There is then the problem of coma, which becomes significant as the focal ratio is reduced – its effect being inversely proportional to the square of the focal ratio. This becomes very significant when the focal ratio drops to four and such a Newtonian would need to be used with a coma corrector such as those produced by Baader or TeleVue. To reduce the vignetting in the corners of the frame, the secondary mirror needs to be quite large, but for wide field imaging the reduction in the micro-contrast of the image due to using a large secondary mirror is not important. The effect will be to reduce the theoretical resolution by about a factor of two, which, in this application, has no real relevance. What is important is to minimise the vignetting of the field of view when the telescope is to be used with a full frame DSLR. A drop in brightness down to 70 per cent at the edge of the field is generally regarded as being suitable, with the use of flat frames easily able to correct for this if necessary. In order to allow the sensor of a DSLR or large sensor CCD camera to reach the image plane, this has to be some distance (~80 mm) from the telescope tube.

Newtonian Based Astrographs

In the same way that field flatteners can be used with refractors to provide a wide, flat field suitable for use with large sized CCD cameras or full frame DSLRs, so can somewhat more complex optical correctors be used with low focal ratio Newtonian telescopes. Such systems reduce exposure times and give wide fields of view, so that a Newtonian with corrector lens can make a superb astrograph, that is, a telescope optimised for imaging – in this case wide field imaging.

In the UK, Orion Optics manufacture a range of 8 to 16 inch 'AG' (AstroGraph) telescopes using a Newtonian design allied to a 'Corrected Field Flattener', a four-element lens based on a design by Harmer Wynne. The f/3.8 mirrors are hand figured to 1/8th wavelength and are provided with sufficient 'back focus' to allow CCD cameras to come to focus even when, as usual, a filter wheel is in the optical path. Astrosysteme Austria (ASA) produce a range of 8 to 20 inch Newtonian based astrographs (N-series) using a parabolic primary allied to a corrector lens assembly designed by Phillip Keller. The mirrors, figured to 1/7th wavelength, have a focal ratio of f/3.6, so allowing short exposures to be made, and have a usable image plane diameter of 50 mm, so allowing the use of full frame DSLRs for imaging. They also produce an H-series astrograph which uses a hyperbolic mirror and corrector lens assembly to give the even shorter focal ratio of f/2.8 and an improved stellar image quality across its 60 mm flat field. The author's image of M33, described in detail in Chapter 21 was taken using their 8 inch Newtonian astrograph and shows superb stellar images right into the corners of the field of view.

Many companies sell 8 inch, f/4 Newtonians optimised for wide field imaging but these must be used with a coma corrector. They will be able to cover an APSC sized sensor, but not a full frame sensor.

The Cassegrain Telescope

The classical Cassegrain telescope utilises a parabolic primary mirror and a small, convex, hyperbolic secondary mirror inside the focus of the primary which directs the light cone through a central aperture in the primary to a focal point beyond. As the light path is folded within the telescope tube, the instrument is quite short in relation to its focal length. The convex secondary mirror multiplies the focal length by what is termed the secondary magnification, M, which is the focal length of the system divided by the focal length of the primary. One result of a configuration with a high secondary magnification (to give a compact system) is to give significant curvature of field.

Variants of the Cassegrain Design

Two variants of the classical Cassegrain are used by amateur astronomers: the Dall–Kirkham and Ritchéy–Cretien. The Dall–Kirkham uses a prolate ellipsoid primary allied to a spherical secondary which is easier to fabricate than the hyperbolic secondary of the classical Cassegrain. It does, however, suffer from greater off-axis coma making it only suitable when large fields of view are not required or a corrector lens is included in the optical path. Superb examples are made by Takahashi in their Mewlon range of 210, 250 and 300 mm aperture telescopes. Optimised Dall–Kirkham telescopes are also available such as the Orion Optics (UK) ODK, f/6.8, range of 10 to 16 inch telescopes. These include an integrally fitted corrector lens to provide a wide flat field for visual and photographic use.

Another variant, the Richey–Crétien design, is now used in virtually all large professional telescopes, including the Hubble Space Telescope. It uses two hyperbolic mirror surfaces which are chosen to eliminate coma, so allowing a relatively large field of view, and are thus highly suitable for astroimaging. The spot sizes of stars away from the axis get larger but retain circular shapes, which are more pleasing than comatic images. Until recently, Richey–Crétien designs have been very expensive – even with small apertures – but modern computer controlled mirror grinding technology has enabled them to become cost competitive with other designs. They have relatively large secondary mirrors, so reducing their micro-contrast, but are ideal for wide field, deep sky astrophotography.

Catadioptric Telescopes

The term simply applies to telescopes that use both mirrors and lenses to form the image. One of the two most common designs, called the Schmidt–Cassegrain, is a development and combination of two other telescope designs. The first is the Cassegrain telescope and the second is the Schmidt Camera.

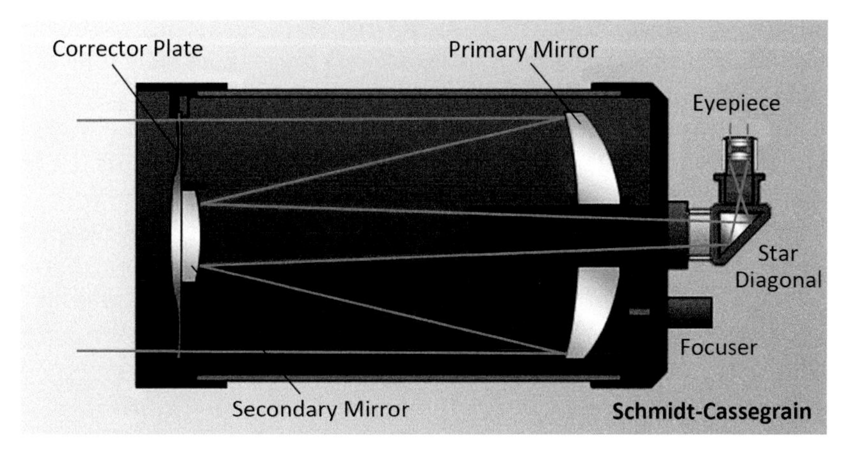

Figure A.12 The optical design of the Schmidt–Cassegrain
(Image: Starizona)

The Schmidt–Cassegrain Telescope

The Schmidt–Cassegrain telescope, or SCT, has become exceptionally popular with amateur astronomers over the past 40 years or so, since Celestron introduced the highly popular 'Celestron 8', 8 inch aperture, f/10, SCT in 1970. A wide range of SCTs are now made by Celestron and Meade. As their name implies, they use the Cassegrain configuration but use a spherical primary mirror allied to a Schmidt corrector plate to correct for spherical aberration that it would produce. In the configuration used in commercial SCTs, the corrector plate also supports the secondary mirror, so there will be no spider diffraction effects. This configuration gives a very compact design – one of the SCTs much prized assets – but does give rise to quite significant curvature of field. This is no real problem for visual observing, but for imaging use, a focal ratio reducing and field flattening corrector lens is added near to the focus. This also reduces the focal ratio from f/10 to f/6.3, so exposure times will be less.

It is vital that the interior of the baffle tube is very well blackened, as otherwise internal reflections from objects such as bright planets or stars within or just outside the field of view can give rise to circular arcs of light. One interesting point is that the corrector plate refracts on-axis light falling close to its edge away from the optical axis. If these rays are to be intercepted by the primary mirror, it must be somewhat larger in diameter than the corrector plate. For a standard 200 mm, f/10 design, which has a corrector plate 200 mm in aperture, the primary mirror needs to have a diameter of 201 mm – not too significant – to give full illumination on-axis. However, if off axial rays that would illuminate the outer parts of the field of view are considered, a primary mirror of some 211 mm diameter would be required if this is not to be an additional cause of vignetting. Looking in detail at the specifications of the Celestron and Meade 8 inch SCTs shows that while Meade use oversize mirrors, Celestron do not. Celestron SCTs will thus suffer a touch more vignetting of the field of view than those made by Meade.

In the majority of SCT designs the primary mirror has a focal ratio of f/2 and the secondary has a magnification ratio of 5, giving the overall focal ratio of f/10. However, Celestron produce one SCT having an aperture of 9.25 inches (235 mm) with a primary of f/2.5 and a secondary having a magnification of 4 – so giving the same overall focal ratio. At the expense of a longer tube than would otherwise be required, this design gives a flatter field of view than the standard design and shows less coma and off-axis astigmatism. It is sometimes said to be the SCT for those who do not like SCTs!

Mirror Shift

In the majority of Schmidt–Cassegrains, the image is focused by moving the primary mirror away or towards the secondary rather than having a focuser at the rear of the telescope. The mirror thus has to slide up and down a track and, as it slides, has a tendency to shift very slightly in its orientation with respect to the optical axis of the telescope. This causes the image to jump within the field of view. When visually observing relatively wide fields this is no real problem, but when attempting to focus an image of a planet onto the small sensor of a webcam for planetary imaging, this can be a problem. This 'mirror flop' could also ruin a long exposure image as the mirror might move as orientation of the telescope tube changes over time. To overcome the first of these problems, a focuser can be added to the rear of the tube assembly so that, once approximate focus is made using movement of the primary mirror, final focus is achieved using the focuser. The SCTs recently made by both companies have far less mirror shift than earlier models, but it is still present to some extent and, in some of the very newest designs, the mirror position can be locked in position once focus is found.

Cool Down Times

As one can see looking at the optical design, the light traverses the optical tube assembly three times. It is thus vital that before any serious observations begin the air inside the closed tube has had time to cool down to the outside temperature. If not, the image quality will be seriously degraded with bloated star images. (The way to check on progress towards this goal is to observe the out of focus image of a bright star: if the telescope is still cooling down, it will appear to be 'bleeding' with streamers extending from the edge of the Airy pattern.)

Two obvious aids to achieving thermal stability are to keep the telescope in a cool location, such as a garage, which will be closer to the outside temperature, and to bring the telescope out into the open air an hour or so before observations are intended to begin. To speed up the process, it is possible to purchase a blower which fits in place of a 2 inch eyepiece and which injects filtered air down a tube towards the secondary mirror.

Figure A.13 The author's 9.25 inch Celestron Schmidt–Cassegrain equipped with a Celestron Dew Shield, Tele Vue Starbeam finder, Starlight Instruments Crayford focuser and (right) a resistor based dew heater surrounding the corrector plate

Dewing of the Corrector Plate

One problem that observers will often find when using a SCT is that, after observing for an hour or so (sometimes less in very humid conditions), the image quality will deteriorate and one will find that dew has formed on the corrector plate. This is because, as the telescope faces the sky, the corrector plate radiates heat away and can reach the dew point temperature, allowing the water vapour in the atmosphere to form a thin covering film. Under very cold conditions one might even (as I have) find frost, not dew, covering the corrector plate!

The simplest way of staving off the time when the corrector dews up is to extend the telescope tube with what is called a 'dew shield'. Its length extending beyond the corrector plate should be at least one and a half times its diameter. A corrector plate without a dew shield is 'seeing' much more of the sky and so cools more quickly than when a dew tube restricts the area of sky visible to it. Many commercial dew shields lie flat for easy storage and are simply wrapped round the tube assembly and fixed with Velcro fastenings. It should be pointed out that a dew tube should have a matt black or black felt lined interior to help prevent light scatter.

If the air is not too humid, a simple dew tube may be all that is required, but often, even when one is used, the corrector plate will eventually dew up. What can one do then short of deciding that it was time to end your observing session? Never attempt to wipe it away, one could easily remove the coatings on its surface which are so important! One possibility is to warm the dewed corrector plate with a portable hair dryer running off a 12 volt supply. (One should not consider using a mains dryer under humid conditions!) This takes only a few seconds and will not significantly alter the corrector's shape.

A better solution is to continuously apply a little heat to the periphery of the corrector plate. This can be done by using a dew shield that incorporates an integral heating element (such as those made by Astrozap for Meade and Celestron SCTs) or by wrapping a heating element around the telescope tube before adding the dew tube. These do use a significant current and a separate high capacity battery should be used for their use. Kendrick, who pioneered this procedure, is among the several manufacturers who provide the dew heating strips and the controllers that can be used to control the amount of heat emitted by them. Do-it-yourself designs for both the heating strip and controller are available on the web. Even under conditions where dewing up is not expected, it is very sensible to add a dew tube. It will act to prevent unwanted stray light falling onto both the corrector plate and the interior of the optical tube which may be scattered onto the camera sensor, so reducing the overall image contrast. It may also prevent unwanted finger marks on the corrector plate!

There can also be a problem when bringing a telescope tube back into the house after an observing session, as it can then also dew up. This will clear as the corrector plate warms up, but it is best to bring it into a cool room and not place a cover (which will trap the water vapour) onto the front of the telescope before it has cleared.

Wide Field Imaging with the Celestron Fastar® Capability

Many of Celstron's latest optical tube assemblies and some earlier ones allow for the secondary reflector to be removed so that a corrective lens assembly can be put in its place, to correct for spherical aberration, coma, off-axis astigmatism and field curvature. A CCD camera can then be mounted on the assembly to provide a very fast, f2, system for wide field imaging. Starizona produce a range of Fastar® compatible correction lenses to which can be attached a variety of CCD cameras. QHY manufacture a range of CCD cameras, such as the QHY8L describes in Chapter 19, having a small circular form and these are perfect for use with them.

Celestron's Rowe–Ackermann Schmidt Astrograph

This is a fast, f/2.2, 11 inch catadioptric system which has a focal length of just 620 mm and offers a full 70 mm optimised image circle. The design uses four elements of rare-earth glass and gives images free of chromatic aberration, coma and field curvature, along with minimal vignetting. A fan is incorporated to reduce cool down times using filtered air.

Improving the Schmidt–Cassegrain

When the SCTs were first designed, 1.25 inch eyepieces with their limited field of view were being used, so the off-axis image quality, which is limited by coma and (for imaging) flatness of field, were not as important as now. To make their telescopes better suited for wide field visual observing and imaging, both Meade and Celestron have now updated their designs with the aim of reducing off-axis coma to improve the quality of star images towards the edge of the field of view and also, particularly in the case of the Celestron design, to make the field flatter. The former actually makes more stars visible as their images are 'tighter' and the latter has become important with the large imaging sensors now being employed in DSLRs and CCD cameras. The two companies have, however, tackled the problem in different ways.

In the Meade ACF – ACF standing for 'Advanced Coma Free' – series of optical tubes, the secondary mirror, rather than having a spherical surface as in the standard design, is given a hyperbolic surface as used in the Richey–Crétian (RC) design. The use of a different corrector plate than in the SC design coupled with the spherical primary mirror gives the same effect as a hyperbolic primary – also used in an RC design. So the resultant optical configuration gives a similar result to an RC telescope which, as well as essentially eliminating coma over the field of view, helps to flatten the image plane so allowing larger CCD sensors to be used for imaging. To eliminate the problem of mirror shift when focusing, Meade have made available a 'Zero image shift microfocuser' that fits between the base of the mirror assembly and the eyepiece. Having neared focus by adjusting the primary mirror position and locking it in position – so there will be no further mirror movement when imaging – the fine focus can then be achieved with the microfocuser, whose position is controlled by the telescope control handset. This is provided as part of the 16 inch system and is an optional extra for the smaller tube assemblies.

A different approach has been taken by Celestron, who have left the primary and secondary mirrors spherical and utilise a standard SCT corrector plate. To achieve the desired objectives of eliminating the off-axis coma and significantly flattening the image plane, Celestron have, instead, incorporated a dedicated optical assembly built into the baffle tube in what they call an 'Aplanic Schmidt–Cassegrain'. Their two objectives have been well achieved, and the tube assemblies incorporate a number of new features as well: flexible tension clutches hold the mirror in place and prevent image shift as the telescope orientation changes, so keeping the image centred in the eyepiece or sensor, while micro-meshed filtered cooling vents located on the rear cell allow hot air to be released from behind the primary mirror. Dedicated reducer lenses for each sized telescope are available to reduce the effective focal ratio to f/7, so giving shorter imaging exposures and wider fields of view. The EdgeHD optical tubes remain Fastar® compatible for ultra-fast f/2 wide field imaging so each tube has been fitted with a removable secondary mirror to allow the Fastar® optics to be installed.

The Maksutov–Cassegrain

As the aspheric surfaces of the Schmidt–Cassegrain corrector plate made it difficult to manufacture (until a neat trick was invented by Celestron), opticians tried to design a corrector plate that would correct for the spherical aberration caused by using a spherical primary mirror but which would only incorporate spherical surfaces. The solution, which was first published by the Russian Dmitri Maksutov, was

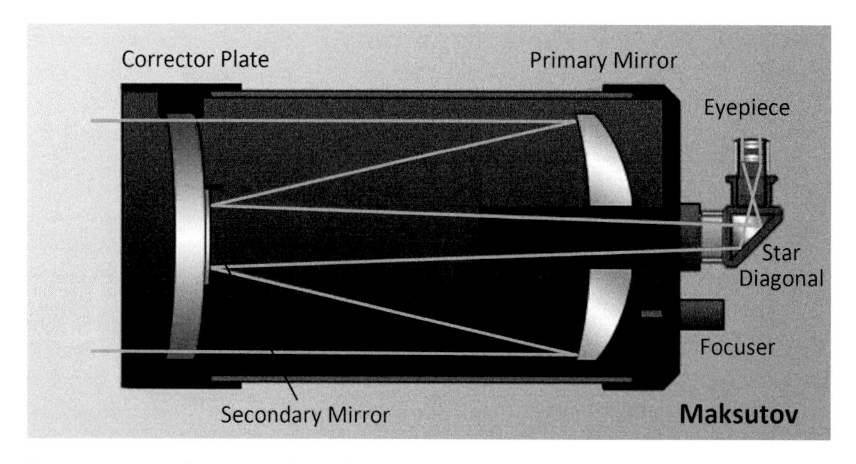

Figure A.14 The optical configuration of a Maksutov Telescope, in this case a Rutten–Maksutov
(Image: Starizona)

to use a 'meniscus corrector', which is a strongly curved negative lens having a very low power. Such a lens introduces spherical aberration of opposite power to that of the primary mirror and so largely eliminates it. The resultant telescope design has become known as the Maksutov–Cassegrain or, more normally, a Maksutov. Maksutovs are very widely used in telescopes of 90 to 127 mm aperture, but the corrector plate becomes very thick and heavy for larger apertures with few being produced with apertures greater that about 175 mm.

In 1959, John Gregory published a Maksutov design in which the secondary mirror was simply an aluminised spot at the centre of the inner surface of the corrector plate so making it very cost effective. This f/15 design is known as a Gregory–Maksutov. An alternative to Gregory's design is to replace the aluminised spot with a separate convex mirror. This, more complex, design was introduced by Harrie Rutten and so is often called a Ru-Mak. It gives the optical designer an additional degree of freedom as this mirror can have a different radius of curvature than the internal surface of the meniscus corrector. This extra degree of freedom allows the Maksutov to provide better off-axis images and gives a flatter focal plane, though the optical tube is longer than in the Gregory–Maksutov. In some designs the effective focal ratio is reduced to f/10 so allowing for a wider field of view.

The Gregory–Maksutovs have been made in very large numbers by Meade and are now available with apertures of 90 and 125 mm and are used in their very popular ETX range of telescopes. Celestron also produce 90 and 127 mm Maksutovs as part of their SLT range.

The Russian company Intes-Micro produce several models ranging in aperture from 5 to 10 inches under the Alter brand name. They have superb optics with wavefront errors better that 1/6th λ and hence produce excellent images, but their cost is substantially more than equivalent aperture SCTs. The Alter range have internally baffled tube assemblies and air outlets surrounding the corrector plate to allow the interior to cool down more quickly. I own an Alter 500 (5 inches aperture) Ru-Mak and it is one of my favourite 'get up and go' telescopes.

Telescopes with Sub-aperture Correctors

Though the vast majority of catadioptric telescopes use full aperture correcting plate, it is perfectly feasible to either place a correcting element between the primary and secondary mirrors – the light passing through the element twice – or to include a lens assembly within the baffle tube. There are a number of such designs, two of which are manufactured by Vixen. In their VMC (Vixen Maksutov–Cassegrain) series of 95, 110, 200 and 260 mm aperture telescopes, a meniscus corrector plate is mounted in the light path in front of the secondary mirror, so this design could also be called a sub-aperture Maksutov. The dewing problems associated with full aperture corrector plates are avoided and, being an open tube assembly, the telescope will cool down to the ambient air temperature more quickly. (I would still advise the use of a dew tube to minimise the stray light that could enter the telescope tube.) However, bright stars will show diffraction spikes as the secondary assembly is supported by a spider. Primary mirrors of ~f/2.5 are used with a secondary magnification of ~4, giving overall focal ratios of 10 to 11. This is typically less that that of Maksutovs of the same aperture, so giving well corrected, somewhat wider fields of view. In the case of the VMC200L, the primary mirror is fixed – so cannot shift during long CCD exposures – and focusing is carried out with the use of a rack and pinion focuser behind the primary mirror so that there will be no image shift when focusing.

Vixen also manufacture the VC200L reflector, which is an 8 inch, f/9, highly corrected telescope ideal for astroimaging use. The design features a high precision sixth order aspherical primary mirror, a convex secondary mirror and a triplet corrector lens mounted within the baffle tube. This complex design provides point like star images (smaller than 15 microns) out to the edge of a full frame (36 mm × 24 mm) CCD camera and boasts a 42 mm fully illuminated field. A dedicated, f/6.4, focal reducer is available to reduce imaging exposure times. Like the VMC200L it also has a fixed primary mirror and rack and pinion focuser, with their attendant advantages. The central obstruction of 40 per cent makes it somewhat less suitable for planetary imaging, but as it is primarily designed for wide field imaging where this is of no consequence.

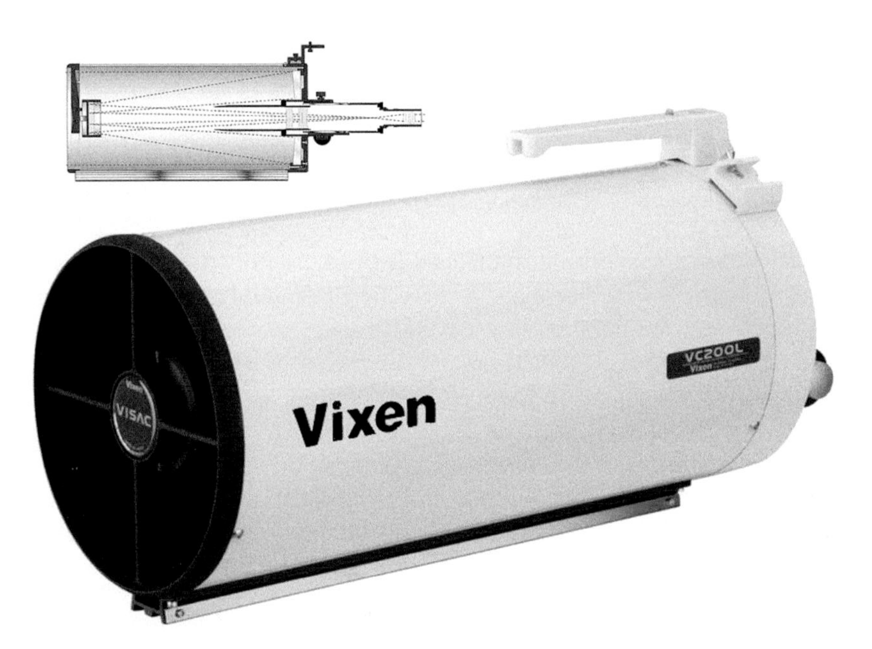

Figure A.15 The 8 inch, f/9 Vixen VC200L showing the three element corrective optics mounted within the baffle tube
(Images: Vixen Optics)

Appendix B
Telescope Mounts

No imaging system can be used sensibly unless it is supported by a mount that is sufficiently sturdy to hold it steady. As telescope apertures increase, it may well be that the cost of a suitable mount will exceed that of the telescope tube assembly, but this is a price worth paying as a poor mount will cause endless frustration when imaging. In this case, a really solid mount is of prime importance and some authors state that one should reduce the nominal load capacity of a mount for this use.

Mounts come in two basic types: altitude/azimuth (alt/az) or equatorial. Before the advent of computer controlled drive systems most mounts were equatorial. The reason was simple. Once an equatorial mount has been aligned on the North Celestial Pole and, say, a Messier object is slewed to, a drive – at a fixed sidereal rate – need only be applied to the right ascension (RA) axis to track it across the sky. Thus a simple electronic controller, based on a crystal oscillator to give an accurate time base, could be used. In contrast, the alt/az mounts need to be driven in two axes at variable rates in order to track an object across the sky.

Alt/az mounts suffer from two problems. The first is that they find it almost impossible to track an object as it goes close to the zenith, and the second is that alt/az mounts suffer from a problem called 'field rotation', which means that over time the orientation of an object will rotate within the field of view. For visual observing with an alt/az mount this is no problem, but when astroimaging, the allowed exposure times will be limited and, ironically, the greatest rate of field rotation is when the object passes through the meridian – when it is highest in the sky and so best placed for imaging! Astroimagers will thus tend to use equatorial mounts, though a program such as *Deep Sky Stacker* can be used to stack a number of short exposure images, having first corrected for the rotation between the individual images.

Altitude/Azimuth Mounts

Besides the Dobsonian mount, which is unlikely to be used for astroimaging, there are three ways to achieve an alt/az mount. Two of these are fork mounted configurations: smaller telescopes will often be mounted on a singlefork mount, whilst larger but compact telescopes such as Schmidt–Cassegrains (SCTs) may use a dual fork mount. The great advantage of a fork mount is that there is no need to counterbalance the telescope, so keeping the total weight of the moving parts down and thus requiring a lighter mount and tripod. In the case of single fork telescopes it may well be impossible to observe towards the zenith to avoid the telescope tube fouling the mount. With dual fork mounts, visual observing may be possible near the zenith, but there may be insufficient clearance to allow imaging cameras of any size to be used. It is possible to purchase equatorial wedges which fit between the tripod and mount so that they can be converted into equatorial fork mount.

Counterbalanced Alt/Az Mounts

A third design needs to be used for heavier and longer tube assemblies such as medium sized refractors and Newtonians. These employ a counterbalanced system where the telescope lies on one side of the mount and a balancing counterweight is located on the opposite side. The mount has to thus support a greater weight than the telescope tube alone. An interesting point is that the counterweight can be a second telescope, so one could employ, say, an 80 mm refractor balanced by a 127 mm Maksutov to provide both wide fields of view for clusters and higher magnification views of the planets. For solar imaging, one could mount an 80 mm refractor with a Baader white-light solar filter or Herschel wedge counterbalanced by a 60 mm H-alpha solar telescope. An important point to note is that alt/az mounts cannot be autoguided so that relatively short exposures are required.

The SkyWatcher AZ-EQ6 GT Alt/Az or Equatorial Mount

This type of dual mount seems to be the 'in' thing at the present time and, perhaps not surprisingly, SkyWatcher have produced an excellent version called the AZ-EQ6 GT, which can support 18 kg for imaging use. In alt/az mode the counterweights can be replaced with a second telescope so greatly increasing the telescope load capacity. It uses a belt drive system to give very smooth tracking so is ideal when autoguiding is to be used when in equatorial mode.

The iOptron Minitowers

iOptron produce three versions of the Minitower: the Minitower V2.0, the Minitower Pro and the newer AZ Mount Pro. The earlier ones look similar but the Minitower Pro employs a sturdier tripod and enhanced bearings to support a payload of 33 pounds rather, than the 25 pounds of the Minitower V2.0, and can support, for example, Schmidt–Cassegrain telescopes of up to 9.25 inches. The mounts have an integral GPS module to set the accurate location and time.

The AZ Mount Pro, having a similar payload capacity as the Minitower Pro, features a 'level and go' intuitive GOTO setup that uses a built-in precision level

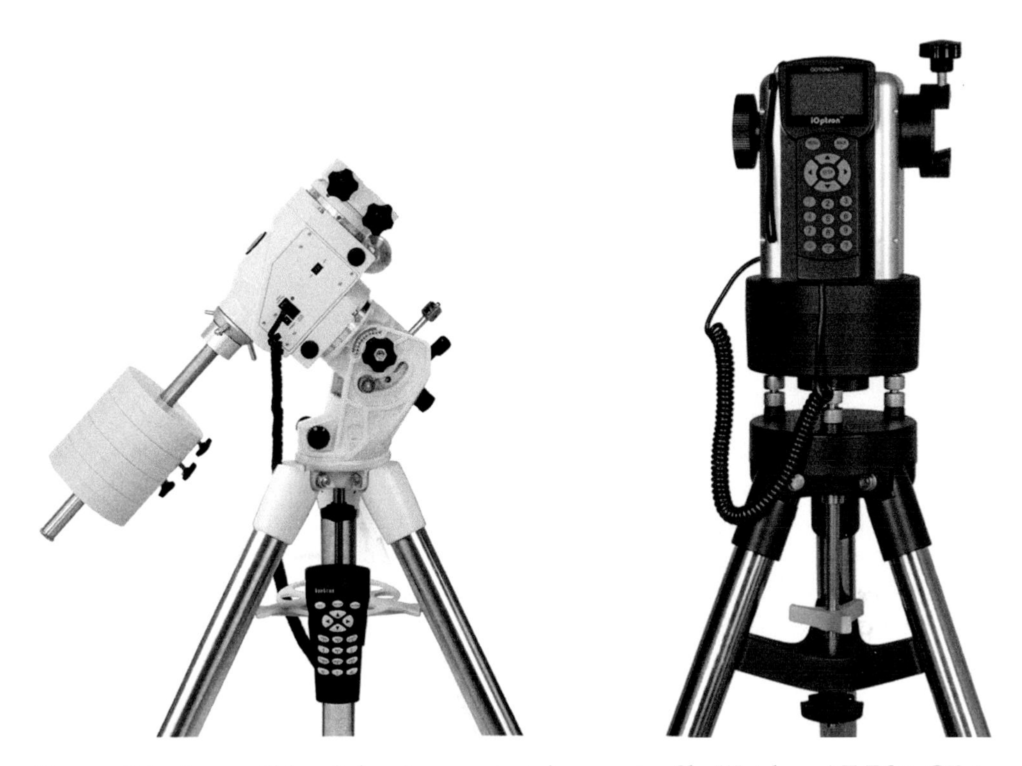

Figure B.1 Convertible alt/az to equatorial mounts. SkyWatcher AZ-EQ6 GT in equatorial mode and iOptron Minitower Pro in alt/az mode
(Images: SkyWatcher and iOptron)

indicator with GPS position and angular detection. The mount will automatically slew to an easily identifiable bright object in the night sky and then the user need only confirm that a bright object is in the centre of the view, to activate 'go-to' slewing to objects which are then tracked with a zero backlash drive system. Uniquely, an onboard rechargeable lithium ion battery is included for up to 10 hours of use.

The Equatorial Mount

These are, of course, the mount of choice for astroimagers as there is no frame rotation and they can be autoguided so allowing for longer exposures. An equatorial mount should be sturdy enough to easily support the weight of the imaging telescope and its camera, often with the added weight of an additional guide scope and camera. It is generally recommended that the load rating of a mount should be derated by 25 per cent if it is to be used for astroimaging. This is particularly important when large refractors are used as these have the considerable weights of the objective at the outer end of the telescope tube and that of the camera (particularly if it is a cooled CCD camera and filter wheel) at the other. This gives it a considerably greater moment of inertia than that of a similar weight, compact, telescope such as a Schmidt–Cassegrain and will require a somewhat more rigid, and hence heavier, mount to be used successfully.

The full name of the type of equatorial mount most used by amateur astronomers is 'German equatorial mount' or GEM for short. Mounted on a tripod is an equatorial head supporting a short tube called a 'polar axis' which has to point towards the North Celestial Pole. This is driven by a motor and drive system so that it rotates once each sidereal day. (Note: the Moon moves quite rapidly across the sky so needs to be followed at a somewhat different rate. Many equatorial mounts have a 'lunar' drive rate as well as a sidereal rate.) In many mounts, a small telescope, the polar scope, is mounted within and aligned along the polar axis to enable accurate alignment on the North Celestial Pole.

A real consideration is how well the telescope tracks the object across the sky – essentially in right ascension with, hopefully, only minor corrections being made in declination. Manufacturers usually quote that the mount will track to a given accuracy (say, of +/− 10 arc seconds, which would be a pretty good specification); however, if stellar images are 3 arc seconds in size this will not allow for sharp, rounded, stars when long exposures are made. The error in the tracking is usually down to the precision with which the worm and gear that drives the mount in right ascension is made. Mounts with larger diameter gear wheels (which are usually more expensive) will tend to give smaller tracking errors. For example, the Losmandy G11, which uses a 5.6 inch diameter, 360 tooth, worm drive, has a quoted tracking error of +/− 10 arc seconds, while the smaller GM8 uses a 2.8 inch diameter, 180 tooth, gear and has a quoted tracking error of twice that. In the case of these Losmandy mounts, a French Company Optique & Vision (www.ovision.com) can supply high precision worms which reduce the tracking error of each by a factor of two; thus an autoguiding system, if used, will need to make lesser corrections to the drive.

Periodic Error Correction

The rack that is used to rotate the telescope at sidereal rate is usually driven by a motor driven worm meshing with a worm gear. For each 360° rotation of the worm, the worm gear rotates by one tooth. Depending on how well the worm has been machined, the rotation rate of the worm gear will vary very slightly as the worm rotates giving rise to a small tracking error. However, the effect repeats very closely for each turn of the worm and so this is called a 'periodic error'. Some mounts allow this to be corrected in what is termed, not surprisingly, 'periodic error correction' or PEC. In the case of my Losmandy GM8, the worm rotates once each 11 minutes. Before the start of an observing run where high precision is required, a recording mode is initiated when, whilst observing a bright star, very minor guiding corrections are made using a very slow slew rate – perhaps twice that of the drive rate – to keep the star centred in the field of view. After 11 minutes the 'recording' mode is stopped and thereafter these fine drive corrections will be applied automatically as the worm rotates, thus correcting the periodic error. PEC is not normally recommended if autoguiding is to be used as then the two correction systems might fight each other.

When the telescope is being autoguided, the overall periodic error is not that important. Far more so is how smooth the mount is tracking, even if the rates vary somewhat. One problem that is found in some mounts is that the grease used in the bearings is not that high in quality and, in this case a complete strip down and replacement with high quality grease can make a substantial difference. If there is stiction in the system, no matter how good the guide commands might be, the mount will not respond accurately. Celestron recommend Klubertemp –40 to +260 grease or its equivalent, such as Mobil 1 Synthetic Grease.

No matter how well a worm and gear are manufactured, there is bound to be some backlash, as if not, the drive would seize up. Avalon Instruments, an Italian company, have avoided this problem by dispensing with a worm and gear and replacing them with a four stage reduction belt drive system that is backlash free. They do not claim very low tracking errors but the drive system is very smooth, so that in conjunction with an autoguiding system, the tracking is superb. Rowan Astronomy (http://rowanastronomy.com/products.htm) can provide a belt drive conversion kit for the very popular EQ6, NEQ6 and NEQ6 Pro mounts, which will greatly improve their tracking accuracy when autoguiding is used.

Unguided Long Exposures

Perhaps surprisingly, this is possible, and the '10 Micron Astro-technology' series of mounts can achieve unguided exposures of an hour or more with a peak tracking error of less than one arc second. This is achieved by mounting highly-precise absolute encoders on each axis of the mount whose positional measurements are accurate to 1/10 arc second. Their outputs are then included in the servo loop that drives the AC drive motors, thus allowing compensation for the periodic errors in the worm drives. The precision is aided by using a zero backlash belt drive to implement the drive reduction. Before use, a pointing model is built up during twilight (so no imaging time is lost) by aligning on perhaps 10–25 stars across the sky. Slews to objects in the extensive database that contains objects down to 16th magnitude are then good to around 20 arc seconds. The mount allows for an object to be tracked 30 degrees past the meridian, so allowing an observing time of four hours when the objects are highest in the sky. The fast slewing rate of 15 degrees per second can actually achieve a meridian flip in around 20 seconds and still have the object centred in the frame. There is an ST4 guiding port, so autoguiding can be used if desired. This might, for example be required for use with a Schmidt–Cassegrain whose mirror is not locked and so might shift slightly. In this case an off-axis guide camera can be used as discussed in Appendix D on autoguiding.

The electronics are mounted in an easily removable control box with all connections locked. This means that, having removed the control box, the mount can be kept outside (suitably covered), so making an observatory less necessary. One interesting feature is that the system can determine if the telescope system is out of balance and provide instructions as to how to adjust the counterweights or telescope position to easily achieve a balance to within ~2 per cent.

The mounts range from a very portable GM1000 HPS having a 55 lb (25 kg) maximum load capacity through the semi-portable GM2000 HPS (132 lb, 60 kg) and the GM3000 HPS (220 lb, 100 kg) up to the GM4000 HPS mount that can support a total payload of 330 lb (150 kg).

How to Align an Equatorial Mount

Before an equatorial mount can be used, the polar axis must be aligned so that it points towards the North Celestial Pole (NCP). This can largely be done in daylight, so maximising your valuable observing time. There are two steps to this process. The first step is to set the correct altitude for the polar axis. This depends on your latitude. This can be found from a map or can be found on Google Maps by bringing up a map including your location (perhaps by entering the ZIP or postcode). Then, by right clicking on your precise location within the map and dropping down and clicking on the 'what's here?' line of the menu that appears, the longitude and latitude appear in the location box above the map in place of the ZIP or postcode.

On the side of the mount is a scale and pointer with adjusting screws to set the polar axis to the correct altitude by setting the pointer to the latitude. Of course, in use, the altitude of the polar axis will not be right unless the equatorial head is level, so the tripod legs should be adjusted before use using a spirit level (in two directions at right angles) to make sure that the base of the head is horizontal. Usefully, some equatorial heads incorporate either a single bubble level or two levels at right angles.

The second step is to point the polar axis towards true north. An obvious way is to use a compass, but this involves knowing the Magnetic Variation at your location. There is a superb website, www.ngdc.noaa.gov/geomag-web/#declination, which will give the deviation (called magnetic declination) from true north at your location. The latitude and longitude can be entered directly or found from your country and nearest city. The date is automatically entered but can be changed for a future date if required, and then by clicking the 'Calculate' box, the current magnetic declination and its rate are given to an accuracy of half a degree. For example, at the time of writing, the magnetic declination at my home location was 1 degree 45 minutes west, so that I should align the polar axis this amount to the right of north as given by the compass. The page also gives the rate of change, which for my location is currently 9 arc minutes per year.

What if you just bring out your scope in the evening? Set the head horizontal and the polar axis to the appropriate altitude as described above, and adjust the azimuth to point the polar axis up towards Polaris. I have made an adapter so that I can mount a green laser pointer in the polar axis tube and simply move the mount so that it points up at Polaris – but not quite. The North Celestial Pole is a little way – 42 arc

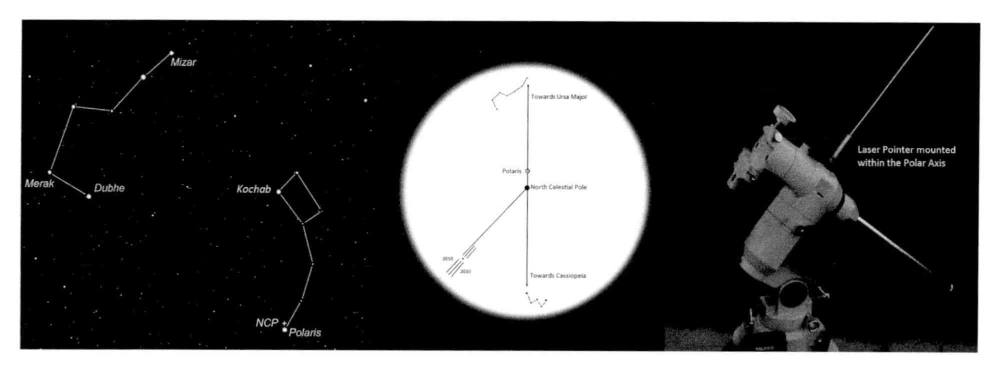

Figure B.2 The position of the NCP towards Kochab, the view through a typical polar telescope and the author's use of a green laser pointer to make an initial alignment to the NCP

minutes or 0.7 degrees – from Polaris towards the star Kochab, the second brightest star after Polaris in Ursa Minor, so the laser needs to be pointed somewhat to the Kochab side of Polaris. If you have done this, then your alignment is pretty good, but for long exposure astrophotography, more accuracy will be needed and I would then use the polar telescope to offset Polaris the right distance and direction away from the NCP. This scope must be rotated so that Polaris is in the right position. Put the point (often a small circle) where Polaris is to be placed so that it is directly opposite Kochab. Some polar telescopes show the direction of the Plough and Cassiopeia and, in this case, the polar scope is rotated so that they lie in the appropriate directions. If not, a planisphere or the *Stellarium* program can be used to find the relative orientation of Kochab to Polaris if you cannot see it at the time of observation. Then when you adjust the mount slightly so that Polaris is within the offset circle, the equatorial axis will be the correct distance away from Polaris towards Kochab, just as required. (Note: it is important that the polar scope is aligned along the rotation axis of the mount. Polaris (or in the daytime a distant object) should be placed in the centre of the polar scope and the mount rotated in RA. When aligned, Polaris will stay centred in the field of view as this is done.)

The North Celestial Pole lies pretty accurately along the line between Polaris and Kochab, so if Kochab can be seen, a useful trick is to remove the telescope and counterweights so the mount can be rotated a full circle, loosen the RA clutch and align the top of the counterweight bar with Polaris and Kochab. The bar should then point to the right of the end star of the Plough's handle. With the clutch locked, the polar scope is rotated so that the line along which Polaris is to be placed is aligned with the counterweight bar. Making sure that the (usual) diagram of the Plough drawn on the reticule is towards the Plough and then, by making adjustment to the mount axes, place Polaris in the appropriate position along the line (or within the circle). A detailed description of how to use the star Kochab to align an equatorial mount has been described by Clay Sherrod at www.weasner.com/etx/ref_guides/polar_align.html.

Equipping a Polar Scope with a Webcam

A polar scope can be used in conjunction with a webcam so that its field of view, overlaid with the marking to show where Polaris is to be located, can be observed on a computer screen, thus saving lying down on what is often damp ground. Video tutorials can be found on the web show how this is done. Many webcams could be used and the one that I happened to acquire is a Microsoft LifeCam VX-5000 having a 640 × 480 pixel sensor and cost me about £15. To mount it on the end of my Astrotrak polar scope I used a redundant 1.25 inch eyepiece barrel. One end of the barrel located nicely over the ring surrounding the lens and was glued into place. The other end of the barrel was rather too wide to fit snugly over the eyepiece of this particular polar scope and so the polar scope's diameter was enlarged slightly by the use of several layers of black insulating tape. It is generally agreed that a good polar scope can align on the North Celestial Pole with an accuracy of ~5 arc minutes.

Using a Computerised Mount for Alignment

A neat alignment trick can be used with a computerised mount. First ensure that the equatorial head is horizontal and align the mount so that polar axis points reasonably close to the NCP. Having put the scope in the 'home' position, the computer will drive the scope to the first alignment star – which may well not even be in the field of view! Instead of using the slow motion drives to centre it in the field of view as you are told to do, adjust the azimuth and elevation of the polar axis to get it approximately right. Then switch off the mount, move back to the home position and start the alignment process over again. It should be a lot better. After, perhaps, a third try it should be pretty well aligned and imaging can begin.

A similar technique has been proposed by Michael A. Covington which can be used when a scope can be 'one star aligned' and then synchronised on. Having approximately aligned the mount, set the telescope in the home position and do a one star align to a bright star well away from Polaris. Use the slow motion drives to centre it in the field of view and 'synchronise to target'. Then slew, under computer control, to Polaris, but this time centre it using the mount adjustments. Switch off the mount, set the telescope in the home position and carry out the same process a second time. This time, when returning to Polaris it should be closer and it should be in the field of view of a wide field eyepiece. Adjust the mount to centre it in the field and repeat the whole process until both Polaris and the star lie at the field centre as one slews between them.

Drift Scan Alignment

Whether or not a guide scope is used, it would be sensible for any astrophotographer who wishes to take long exposure CCD images to set up their equatorial mount as accurately as possible. To get the ultimate precision a technique called drift scan alignment is used. The basic idea is to centre the telescope on a bright star, stop the RA drive and observe the drift across the eyepiece.

Using a CCD Imager to Observe the Telescope Drift

The process can be made a little easier if the main camera has a CCD or DSLR imager:

- Focus the camera on a star close to zero degrees declination and lying due south.
- Once focused, move the star to the west side of the image field.
- Select the lowest slew rate drive speed.
- Set the camera software to take a 105-second exposure and initiate the exposure.
- After the first five seconds have elapsed, press the West button on the telescope keypad to cause the star to move towards the opposite side of the sensor.
- After 55 seconds reverse the telescope direction to drive east. (That is, after 50 seconds of driving.)
- When exposure has finished, stop moving the telescope and observe the downloaded image.

The image should show a V shape; the more open the V, the greater the misalignment in the azimuth position of the polar axis. If the bright spot marking the initial point on the track is lower than the final point of the track, the polar axis is pointing too far west, so make an adjustment to rotate the polar axis towards the east and vice versa. (Depending on the telescope type, it might be the other way round, so if on the following exposure the V has opened out, make the corrections the other way round.) Repeat the process until the V has collapsed to a straight line.

The telescope is now slewed to a star at low elevation either due east or due west. The procedure is carried out as before, but this time it is the altitude of the polar axis that is adjusted.

As the accuracy in both axes improves, taking longer exposures will allow more precision and the drift line can even go off the image sensor before the drive change brings it back on. As when carrying out the adjustment visually, the greatest accuracy will result when a second iteration of the two operations is carried out.

I suspect that any astroimagers who really need to use the drift scan technique will have an autoguiding system using a CCD camera and may well be using *PHD* guiding. (*PHD* guiding can be downloaded for free if another program is used for guiding.) The process is very similar to that just described, but produces a graphical display of the drift, allowing the appropriate corrections to be made very quickly.

Drift Alignment Using *PHD* Guiding

- Mount the CCD camera on the guide telescope and align it so that the chip is orientated along the RA and Dec. axes with the Dec. axis Up/Down.
- To set the RA correction, choose a star close to declination zero that is near the meridian, due south; first calibrate the mount and then start guiding on it using *PHD.*
- Turn on the Graph and select DX/DY rather than RA/Dec. and turn off the Dec. guiding.

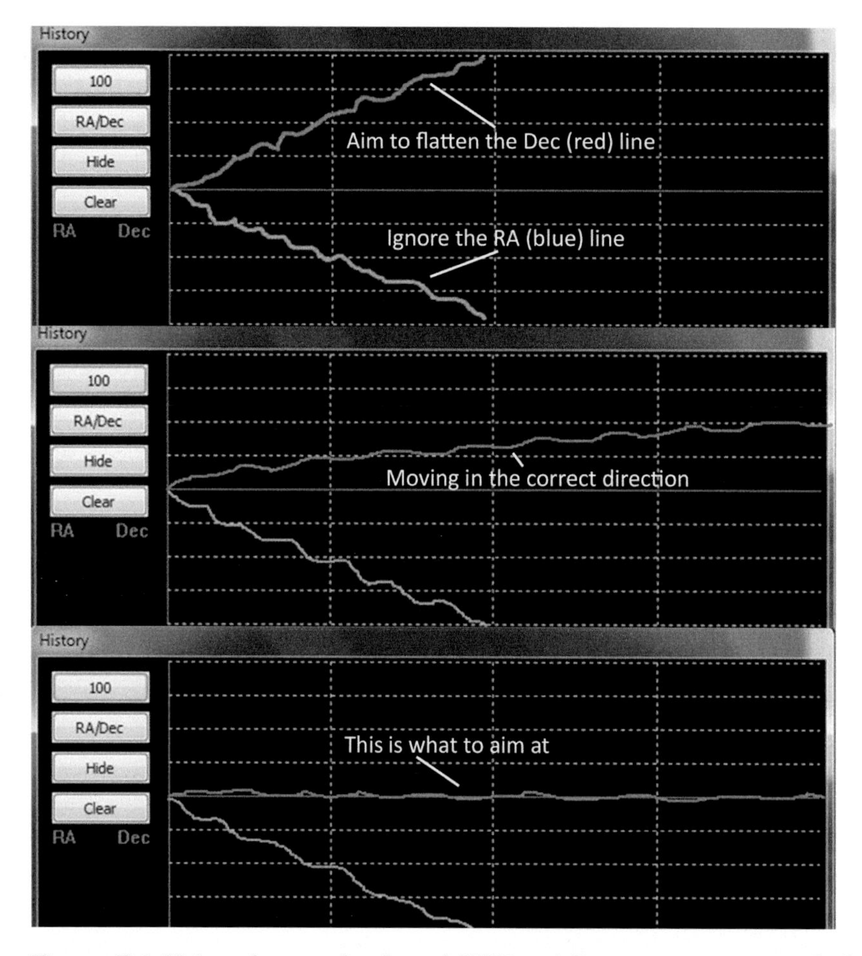

Figure B.3 Using the graph plot of *PHD* guiding to carry out a drift scan polar alignment

- If your mount is perfectly aligned, the DY line (red) will be seen to track horizontally across the graph. Usually it will be seen to drift upwards or downwards. Then make fine adjustments (using the azimuth adjusters) to the azimuth position of the polar axis until the drift is eliminated. The effects of alterations will be immediately visible!
- Now slew the telescope to a star that lies near the horizon due east or west and recalibrate the guiding. Repeat the two previous steps, making sure that the Dec. guiding is off, but this time adjust the elevation of the polar axis when the result of the changes will be seen immediately.

The result of using any of these three techniques should result in a perfectly aligned mount but then very slightly offset the polar axis so that the autoguiding declination corrections are always in the same direction to prevent backlash in the declination gears.

Software Programs to Aid Polar Alignment

There are several programs that will aid polar alignment. These include the freeware programs *PoleAlignMax*, *EQ Align* and *Polar Finder* along with the low cost programs *WCS*, which uses a webcam, and *Alignmaster*, which can be used either with a CCD camera or webcam or an eyepiece with illuminated cross hairs. The latter is quick and easy to use and can achieve a polar alignment to an accuracy of ~1 arc minute. At a cost of 14 euros it is well worth trying but does require, along with the other programs, that the mount can be controlled by computer.

Polar Alignment Using the QHY PoleMaster

This is an innovative new product that uses a lens and camera system that is usually mounted at the front of the hollow polar axis tube to point along the RA axis of the mount. Mounting adaptors are provided for a wide range of equatorial mounts. (If QHY do not provide one for your particular mount, Cyclops Optics (www.cyclopsoptics.com) can provide further adapters.) The objective of the camera and its associated control software is to find the position of the North Celestial Pole and align the RA rotation axis precisely in its direction. A highly sensitivity camera having a field of view of 11 × 8 degrees is used to capture the image of sky around Polaris (so an approximate alignment is required before use). The high sensitivity allows faint stars near Polaris to be imaged, so allowing the software to calculate the position of the NCP.

Polaris is double clicked upon, and an overlay of red circular patches is shown that can be rotated to align on four surrounding stars to confirm that Polaris has been correctly located. One star is then selected by double clicking on it, and the mount is then rotated in RA from its 'home position' by ~20 degrees and the same star clicked upon. The mount is rotated by a further ~20 degrees and the star clicked on again. At this point a green circle appears, passing through the selected star and centred on the North Celestial Pole. The mount is slewed back to its original position in RA and if all is correct, the selected star will be seen to follow round the circle to its original position. When returned to the home position, the program will then place a small green circle on the screen. The mount is then simply adjusted so that Polaris lies within this circle. The polar axis is then aligned with an accuracy that can approach 30 arc seconds (determined by the 30 arc seconds per pixel image scale of the camera). Given such accurate alignment, a single star calibration is usually all that is then required – simply slewing to and then centring on a single bright star and 'synching to target'. The whole process can be achieved in close to five minutes. There are excellent video tutorials for the use of the PoleMaster on 'YouTube'.

I acquired one just as this manuscript was being submitted and found that it worked superbly. The screen gave precise instructions about the various steps that are undertaken. It was amazing to see all the satellites passing through the 11 × 8 degree field of view. I found that the touchpad on my laptop controlling the PoleMaster was a somewhat too sensitive and it was hard to perfectly align the cross hairs onto Polaris and the selected star – both seen within a 'zoomed' window. I shall use an external mouse set to low sensitivity for further alignments. I believe that this device will totally change the way that mounts are aligned in future.

Locating Objects with an Equatorial Mount

In principle, using dials on the mount axes one can arrange the telescope to point directly towards an object given its right ascension and declination and knowing the Sidereal Time (which is used calculate the 'Hour Angle' of the object). I have never done this and do not regard it a particularly rewarding process. As the telescope can be moved in both in declination and right ascension (that is around the polar axis), in many cases it is possible to 'star hop' to the target. But supposing the objects to be observed are somewhat dim, such as, for example, the pair of galaxies M95 and M96 in Leo. If they are found on a star chart, it will be seen that they lie on almost exactly the same declination as the bright star Regulus, Alpha Leonis (+12°). Regulus can be centred in the field of view and the declination axis locked. From the star chart it can be seen that the pair of galaxies are 37 minutes of time in right ascension to the east of Regulus. Using the Hour Angle scale one could, in principle, rotate the telescope in right ascension by this amount to bring them into the field of view of a wide field eyepiece. Again, I do not tend to do this, but simply sweep the telescope slowly eastwards whilst observing though a wide field eyepiece until they appear.

Star Hopping with a Computerised Mount

On the face of it, this seems nonsensical: star hopping is used with non-computerised mounts! But, in fact, it is not quite so stupid as it might appear. If a computerised mount has to slew a long way to find a target object, its precision may only be sufficiently good that an object will only be visible in a wide field, low power, eyepiece. Usually this will be fine and the object can be centred before using a higher power eyepiece if so desired. But if the object has a very small angular diameter and so will appear as a star in a wide field eyepiece, there can be a real problem. One then really needs the object to be within the much smaller field of view of a high power eyepiece so that its true identity can be seen.

An excellent example is the Eskimo (or Clown) Nebula in Gemini. This planetary nebula has an overall size of 48 × 48 arc seconds, but its bright central core is only half this size and cannot be distinguished from a star in a low power eyepiece. So how can it be easily found? The answer is to star hop. It is not that far from the 1.1 magnitude star Pollux, Beta Geminorum, which will usually be found in the bright star catalogue of the controller. First slew to Pollux, centre it in the imaging field and synchronise to it. This will greatly improve the pointing accuracy in this part of the sky. The Eskimo Nebula is only eight degrees away, so the accuracy of most mounts should then be able to place it within the sensor. Its true identity can then be seen and so it can be centred on the sensor.

Figure B.4 Vixen Sphinx, SkyWatcher EQ6, Losmandy GM8 and iOptron ZEQ25-GT equatorial mounts
(Images: Vixen, SkyWatcher, Losmandy and iOptron)

A Survey of Equatorial Mounts

An astronomy 'blog' that is well worth reading is written by the author Rod Mollise – search for 'Uncle Rod's Astro Blog'. The August 2012 contribution contains a very useful survey of many of the equatorial mounts that are currently available. Browsing through his other monthly contributions provides useful comments about the amateur astronomy scene and mini-reviews of many interesting products.

A 'Chinese' Equatorial Mount

iOptron have brought out two interesting variants of the equatorial mount, the ZEQ25-GT and CEM60, which use a 'Z' configuration of telescope and counterweight shaft. They have been called Chinese equatorial mounts and provide a more stable configuration than a standard German equatorial mount. As a result, the ZEQ25-GT allows a mount weighing only 10 lb (4.7 kg) to support a telescope weighing up to 27 lb (12.3 kg). A polar telescope (which is never obstructed) is included along with a 32 channel GPS system to aid setting up. iOptron can also provide a 'PowerWeight™' 8 ampere-hour battery that takes the place of the counterweight to provide a very neat system. The CEM60, which has a maximum payload of 60 lb (27 kg), has a mount weight of just 27 lb (12.3 kg). The head includes two 12 volt power outlets and four USB ports to minimise the cabling associated with, for example, the imaging and guide cameras.

iOptron have also brought out the CEM25-EC, which also has the same payload capacity of 27 lb (12.3 kg) and, like the 10 Micron mount described above, uses high precision optical encoders within the drive servo loop to give a periodic error of just 0.3 arc seconds! It does include an ST4 auto guiding port, but given accurate polar alignment I cannot see that guiding would be required unless very long exposures were required.

Premium Equatorial Mounts

For the really keen amateur astroimager, there are a number of companies that produce very high quality equatorial mounts. Losmandy, Astro-Physics and Takahashi, all of whom have been established for many years, provide a wide range of mounts, including those capable of carrying heavy telescope loads. Losmandy manufacture the GM8, G-11 and Titan mounts – with the latter having an instrument capacity of 100 lb (45 kg). Those made by Astro-Physics Inc. range from the lightweight Mach1 GTO, which is designed for portability but is extremely rigid and has excellent tracking ability, through the 900GTO and 1600GTO mounts up to the 'El Capitan' 3600GTO mount, which is conservatively rated at up to 300 lbs (125 kg). A nice feature is that the control electronics unit (having all connections locked in place) can be easily removed from the mount. I am thus happy to leave my Mach1GTO mount permanently set up in my garden protected under a thick 'duvet' and waterproof cover. When autoguiding using the *PHD* guiding software, the graphical output shows that the mount is achieving a 0.6 arc second rms tracking accuracy when used with an 890 mm focal length refractor. Takahashi's mounts range from the lightweight EM-11, which has a 19 lb (9 kg) payload capacity through the EM-200 (39 lb, 18 kg) and EM-400 (83 lb, 38 kg) up to the EM-500 rated at 100 lbs (45 kg) payload capacity. They have peak tracking errors of 5 arc seconds.

There are some new companies producing very attractive mounts such as '10 Micron Astro-technology' whose mounts were described above. Avalon Instruments produce an innovative single fork equatorial mount, the M-Uno. This has abandoned the traditional worm and gear drive in favour of one that uses toothed belts. These provide a very smooth motion and the complete elimination of backlash. The single fork design does not require a long counterweight arm and there is no 'meridian flip issue' as is the case with German equatorial mounts. This means that the mount will track a target from east through the meridian into the west uninterrupted. It is equipped with the SkyWatcher Synscan 'go-to' controller. ASA Astrosysteme Austria produce a range of high performance, direct drive mounts ranging from the DDM60/PRO portable mount (61 lb, 28 kg load capacity), designed for astrophotographers travelling to dark sky sites, through the DDM85 (100 lb, 45 kg) up to the DDM160 (660 lb, 300 kg) observatory class mount.

Software Bisque have recently brought out the Paramount MYT portable mount which can carry telescopes up to 12 inches (0.3m) in aperture with a 50 lb (23 kg) payload capacity. This has a maximum of 7 arc seconds peak-to-peak periodic error before correction. It uses TPoint telescope modelling software to provide a 30 arc second all sky pointing accuracy. One problem, perhaps, for use at dark sky sites is that it requires a 48 volt supply for which a mains power adapter is supplied. Their Paramount MX+ has a 100 lb (45 kg) instrument capacity, while, perhaps the ultimate amateur mount for observatory use, is the Paramount ME II Equatorial Mount capable of handling an instrument weight of up to 240 lbs. It uses belt driven drives to eliminate backlash.

Appendix C
The Effects of the Atmosphere

Sadly, unless we have access to the Hubble Space Telescope, we have to observe the heavens through our atmosphere. This has two effects on the image we observe caused by the atmospheric 'seeing' and 'transparency'.

Atmospheric Seeing

This is the name given to the blurring and twinkling of astronomical objects caused by turbulence in the Earth's atmosphere. There are formal definitions based on the diameter of what is called the 'point spread function' or 'seeing disc' that corresponds to the fuzzy blob seen in a long exposure photograph of a star. At high altitude observatories, such as that at Mauna Kea or La Palma, the seeing disc may be as little as 0.4 arc seconds in diameter, but from ground level it could be as much as 4 arc seconds, though there are locations such as in Florida or Barbados where the seeing can be surprisingly good. This is attested by the wonderful images of Jupiter taken by Damian Peach from Barbados using a technique called 'lucky imaging', which was described in Chapter 13.

Amateur astronomers use the 'Pickering Scale', devised by William H. Pickering, to describe the quality of the seeing during an observing session. This has a scale of from 1 to 10 ranging from 'Very poor' to 'Excellent'. An excellent visual animation of how a star image will appear under different seeing conditions can be found at: www.damianpeach.com/pickering.htm. It is only in the highest four seeing categories that the first ring of the Airy disc can be seen, but even when the seeing is not that good there can be brief moments of clarity when a still patch of air comes between the planet and observer. A simpler five category scale can also be used:

1)	Bad	Boiling image without any sign of diffraction pattern.
2)	Poor	Broken up central disc. Missing or partly missing diffraction rings.
3)	Average	Central disc deformations. Broken diffraction rings.
4)	Good	Light undulations across the diffraction rings.
5)	Excellent	Perfect, near motionless, diffraction pattern.

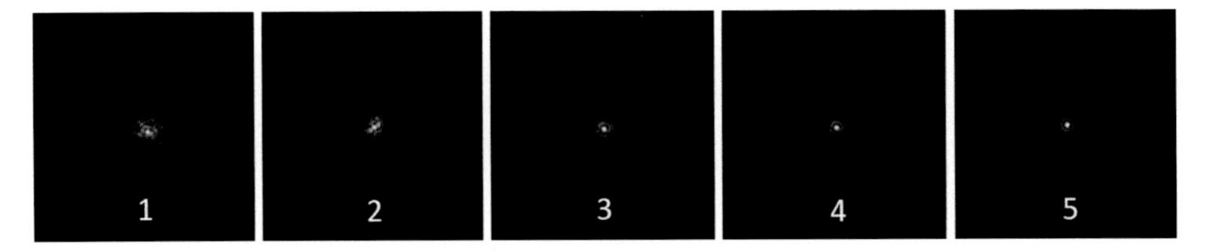

Figure C.1 A five point seeing scale as described in the text

The 'seeing' is caused by small scale fluctuations in the atmosphere both of temperature and density giving rise to turbulence that results in a blurred, moving or scintillating image. This latter aspect manifests itself in the 'twinkling' of stars which are thus a good indicator of how good the seeing is. The fact that the planets are less prone to scintillation indicates that the angular sizes of the atmospheric cells are of the order of arc seconds across.

There are three levels in the atmosphere where turbulence occurs:

1) **Near ground (0 to 100 m).** This is due to convection currents either from the heat of nearby houses or resulting from the cooling of the ground, concrete or tarmac that has been heated during the day. An unvarying terrain such as large areas of grass or water tend to suffer least as they lose their heat more slowly and evenly. Unless the telescope itself has reached ambient temperature, it can cause turbulence in the surrounding air. If mounted on a steel pillar within an observatory, this should be lagged. Run-off roof observatories tend to have better characteristics than a dome.
2) **Central troposphere (100 m to 2 km).** The turbulence at these altitudes is determined largely by the topography and population areas upwind of the observing site. The air downwind of a city or range of hills will contain turbulent eddies which destroy the image quality. It is thus best to have an observing site with the sea or a flat arable landscape for a considerable distance upwind. These help to produce a laminar air flow, resulting in stable images. It is thus not surprising that the world's great optical observatories are located on islands far out to sea or with an ocean upwind of the observing site.
3) **High troposphere (6 to 12 km).** Jet streams in the upper atmosphere can often produce images that appear stable but are devoid of fine detail. Their positions vary with the time of year. They can bring trains of weather systems across a country, so also limiting the numbers of clear nights when one could make observations. Jet stream forecasts for northern Europe and parts of North America can be found at: www.netweather.tv/index.

   The Canadian Weather Office provide a wonderful facility for observers in North America which is a map showing the seeing quality on the five point scale given above for up to 48 hours ahead. This can be found at: www.weatheroffice.gc.ca/astro/index_e.html.

Seeing will be poor when a cold front has passed over, replacing warm air with cold. This gives rise to local convection currents as the ground and buildings lose heat. Conversely, warm air tends to be more stable, particularly when a high pressure area is present and mist or fog has formed. At these times, though the 'transparency' (see below) will be poor, the seeing can be excellent. The formation of cumulus clouds in the afternoon indicates convection in the lower atmosphere, so seeing will tend to be poor for some time after sunset and, as a general rule, will tend to be better in the hours before dawn. High altitude cirrus clouds and light winds often indicate that a night of good seeing is in prospect. The ultimate test is, of course, to observe the image of a bright star and compare it to the videos on Damian Peach's website given above. If the image of the star is defocused, then the patterns of the atmosphere's turbulence passing across the telescope can easily be seen.

In the United States, the states where the seeing tends to be best are those in the south-east, with southern Florida enjoying good seeing throughout much of the year owing to a stable high pressure area and a smooth airflow from the Gulf of Mexico. Arizona and New Mexico fare well too. The more northerly states towards the Canadian border tend to have poorer seeing owing to the presence of the Polar jet stream.

In the United Kingdom, seeing conditions are often good in the south and south-east, particularly near the coast, but further north the seeing is often poor (I can attest to this!). The seeing is usually better during the summer months than in the winter. This is not always the case as in 2012, the jet stream settled over the UK, bringing one of the wettest summers for many years and giving few opportunities for observing and poor seeing when one could.

Atmospheric Transparency

The fact that our atmosphere absorbs most of the infrared and ultraviolet radiation that passes through it is fortunate for us but, happily, the ozone layer that protects us from the ultraviolet only absorbs one or two per cent of the light visible to our eyes. However, even clean air scatters light by a process called Rayleigh Scattering. This affects blue light far more than red, which is why our daytime sky is blue and why the Sun seen close to the horizon is red, as the blue light has been scattered away from the line of sight. Without Rayleigh Scattering, our daytime skies would be black.

At the blue-green wavelengths to which our eyes are most sensitive at night, Rayleigh Scattering reduces a star's brightness by 0.14 magnitude at the zenith with ozone adding another 0.016 magnitude: this is called the zenithal extinction. The extinction obviously increases as a star is observed closer to the horizon as what is called the 'air mass' increases from 1 at the zenith to 2 air masses at 30 degrees elevation and 5.6 air masses at 10 degrees elevation. A star will then appear 0.32 and 0.9 magnitude fainter, respectively. It is obviously best to observe objects as high in the sky as possible.

Sadly, the air is never totally clean, with dust, humidity and emissions from power plants, aircraft and motor vehicles combining to form what are termed 'aerosols'. The result is that, even from a totally dark sky site, a star's brightness will be reduced.

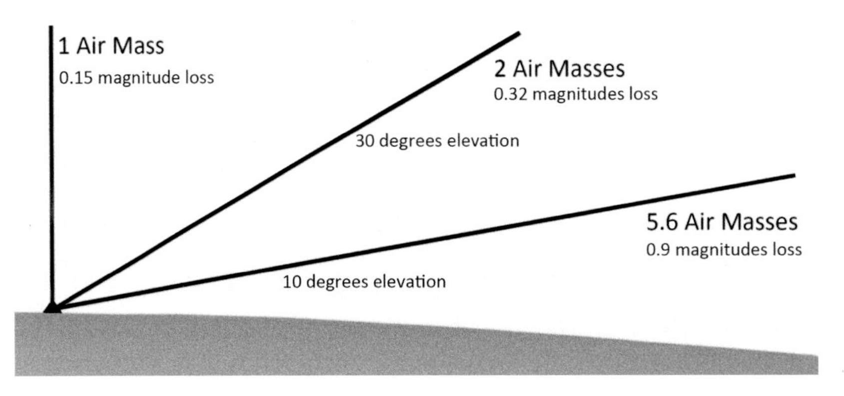

Figure C.2 The effect of air mass on extinction

For example, on a moderately poor night a star that is observed at an elevation of 30 degrees above the horizon will lose 0.8 magnitude and thus only 48 per cent of its light will reach our eyes. (If you could observe from the top of Mauna Kea, the star's brightness will only be reduced by only 0.3 magnitude and 76 per cent of its light will be detected – one reason why the world's largest telescopes are placed atop high mountains.)

The aerosols in the atmosphere have a second effect: that of reflecting light from the ground back to Earth – light pollution. The greater the amount of aerosols in the atmosphere, the greater the amount of light that is reflected back, so helping to mask the, already attenuated, light from the star or deep sky object.

It so happens that the nights when the transparency is very poor, the seeing is usually very good and a calm night in autumn can be really excellent for observing and imaging the planets. Conversely, when the transparency is very good, the seeing can be really bad, so then, should the Moon not intrude, deep sky objects that do not need a high magnification to image are the better targets.

The Canadian Weather Office provide a second facility for observers in North America which is a map showing the sky transparency on a 5 point scale from 'excellent' down to 'very poor' for up to 48 hours ahead. This can be found at: www.weatheroffice.gc.ca/astro/index_e.html.

Airglow

Even from a totally dark site, there is still a further effect which reduces limits the faintest objects that can be seen or imaged. Termed 'airglow' it is light emitted by atoms excited by the Sun's ultraviolet light during daytime. The most prominent colour is the green O III emission line emitted by oxygen in the high atmosphere. Airglow is most prominent during solar maxima when it can reduce the limiting magnitude by 2/3 of a magnitude.

Appendix D
Autoguiding

The outline of the technique is given in Chapter 9 with this appendix describing in more detail both the technique and the guide scopes and guide cameras that are used.

The guide scope has two relevant parameters: the first is its aperture, with larger objectives able to see fainter stars in a given time. For example, an 80 mm guide scope has an aperture 2.56 times greater than that of a 50 mm guide scope and so will be able to detect stars almost exactly one magnitude fainter. So an 80 mm refractor will see fainter stars, but as its focal length will be longer it will have a smaller field of view. The focal length of the guide scope will determine the field of view 'seen' by the guide scope and, obviously, the wider this is, the more likely it is that at least one suitable guide star will be seen. So, to give wide fields of view, a short focal length is best. This, coupled with a large objective, implies that an objective with a low focal ratio would be ideal. The calculation of the optimum specification depends on the density of stars in the direction of view and there is no single answer. However, camera lenses tend to have shorter focal ratios than telescopes and, coupled with reasonably sized objectives giving focal ratios of f/2.8 or even less, might well prove excellent in this task. By providing a wide field of view, it is almost certain that a suitable guide star will be found without having to offset the guide scope from the main telescope's target. This tends to allow a more rigid mounting to be used, so that the guide scope does not move relative to the imaging telescope, which is vital if the guiding is to be accurate.

This idea has been taken up by several manufacturers. Orion Optics (USA) sell a very neat 50 mm objective, 180 mm focal length, f/3.6 guide scope that is essentially a finder scope equipped with a 1.25 inch eyepiece socket. This uses the standard finder dovetail mount that many scopes are equipped with. If not, a dovetail is provided to fix to the imaging scope. Borg offers a 50 mm objective guide scope with a focal length of 250 mm and, even better, a 60 mm scope with a focal length of 228 mm. SBIG offer an exceedingly compact 100 mm focal length, f/2.8 guide scope which is essentially a camera lens provided with a very small and solid mount. As its field of view when used with their ST-I camera is over 2 degrees wide, it is virtually certain that a guide star can be found without having to offset its position relative to the imaging scope. This means that the mounting can be very solid, reducing the problem of differential deflection when the imaging and guide scopes do not remain perfectly aligned. SBIG state that even though

Figure D.1 Orion 50 mm guide scope (top left); Mini Borg 50 mm guide scope (bottom left); and SBIG lens based guide camera (right)

a colour CCD sensor is used in the camera provided as part of the guide package, so reducing its sensitivity, one second exposures can detect 10th magnitude stars.

To some extent the two factors of field of view (reducing with larger aperture guide scopes) and sensitivity (increasing) cancel out. However the tracking accuracy of a longer focal length guide scope will be greater as the image scale is greater, and so the guide software will be able to make finer corrections to the tracking of the mount. I have seen it written that given a 50 mm guide scope, the imaging scope should not have a focal length greater than ~1500 mm. I am using imaging scopes ranging from 400 mm up to 1800 mm and so have purchased an 80 mm, f/5 achromat guide scope having very rigid CNC tube rings so, mounted on a crossbar parallel to the imaging scope, there should be little differential flexture between the guide and imaging scopes. This is a simple achromat and, to tighten the star images, I use a semi-apo filter to remove the extremes of the visible spectrum.

The alternative to mounting the two scopes on a crossbar is to mount the guide scope on top of the imaging scope. This is often done using ring systems similar, but typically larger, than those that used to be used for finder scopes. By adjusting the six screw threads that hold it in place it is possible to offset the guide scope (which can be very useful as seen below), but I am not convinced that these will be sufficiently rigid.

It is sometimes said that such a guide scope equipped with a sensitive guide camera will always find a guide star. This may well be true under dark skies, but I am not so sure that it is true under light polluted skies. To find a suitable guide star it may then be necessary to offset the guide scope somewhat (but not too far as the guiding accuracy will be impaired). An Avalon Instruments X-Guider or SkyWatcher Guidescope Mount as seen in Figure D.2 can be located between the crossbar and guide scope to allow it to be offset in either direction if needed to find a suitably bright guide star.

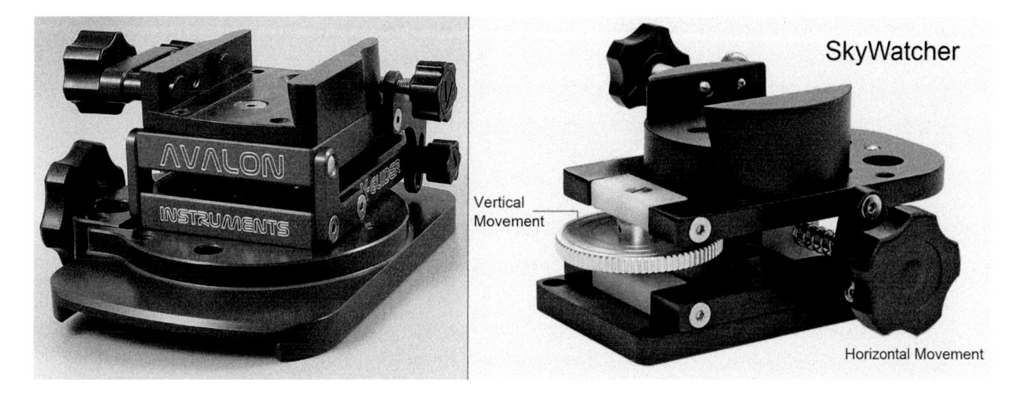

Figure D.2 Avalon Instruments X-Guider and SkyWatcher Guidescope mounts

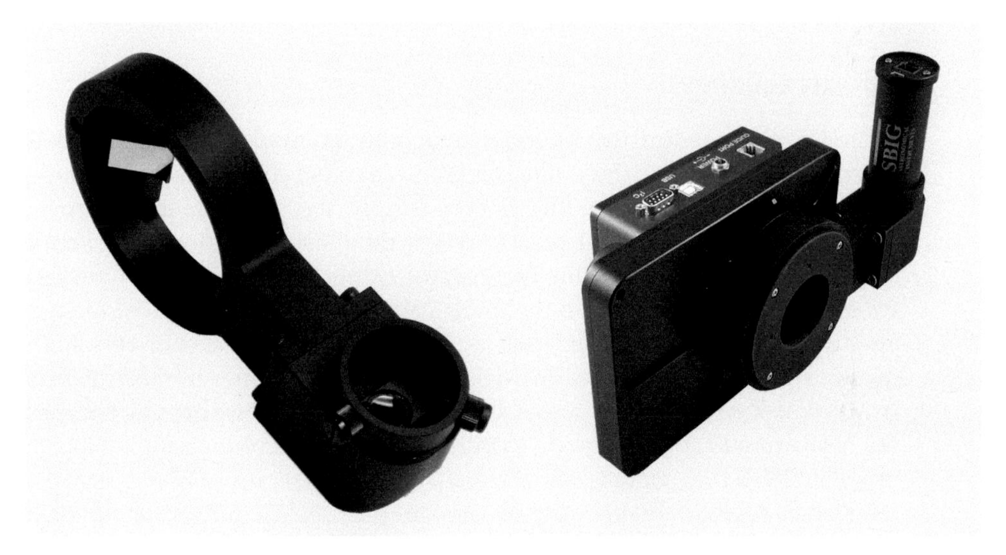

Figure D.3 Off-axis guider for use with SBIG ST-8300 and STF-8300 CCD cameras and ST-i planet cam/guide camera
(Image: SBIG)

Off-axis Guiding

There can be a problem with many Schmidt–Cassegrain telescopes in that the position of the primary mirror is not rigidly located and can sometimes move very slightly, so affecting the pointing of the telescope. Some of the latest designs, which are built with astroimagers in mind, allow the mirror to be locked in position when focus has been found. An alternative approach, as used in the Vixen VC200L for example, is to have a fixed primary mirror and employ a focuser at the rear of the telescope. In these cases, a guide scope system as described above is probably the best solution, but if not, an off-axis guiding system can be used.

This is where some of the light at the edge of the field of view of the main imaging camera is diverted into a separate autoguiding camera. There are two significant advantages to this approach. First, no additional guide scope is required, so reducing both the weight and complexity of the system and, secondly, if the imaging telescope is a Schmidt–Cassegrain, it will correct for any slight movement of the primary mirror during the exposure. Some CCD cameras incorporate a second small CCD immediately adjacent to the imaging CCD; otherwise a small prism or mirror is used to deflect some light from the light path entering the imaging CCD camera and this is focused on a secondary guide camera. These attachments can be rotated so that areas all round the main field of view can be searched, but there can still sometimes be a problem finding a suitable guide star. A secondary problem that can be encountered when a large CCD array is being used is that the images beyond the field edge may well suffer from coma and astigmatism, so giving somewhat misshapen stars that the guiding software may have a problem analysing.

On-axis Guiding

An innovative product has recently come onto the market which enables the main telescope to be used as the guide camera. It is called the Innovations Foresight On-Axis Guider (ONAG®). The idea is very simple: the unit uses a 45 degree dichroic mirror in the telescope light path to reflect the *visible* light (including that from H-alpha emissions) sideways into the imaging camera while the mirror passes the *near infrared* light though and into the guide camera. (The guide camera must not have an infrared blocking filter in front of the CCD!) The mirror is of very high optical quality so as not to introduce any degradation of the captured image and reflects 99 per cent of the visible light into the imaging camera whilst passing ~93 per cent of the near infrared into the guide camera. Two versions of the unit are available: the original ONAG® is suitable for sensors with diagonals up to 28 mm across (such as the Kodak KAF8300) with a larger version, the ONAG® XT, available for use with sensors having diagonals up to 50 mm across. The units are equipped with a movable stage on which to mount the guide camera. This allows a field of diameter 46 mm to be explored (covering over one degree with a typical SCT) to allow a suitable guide star to be found. Although, with a large aperture telescope – say an 11 inch SCT – the field covered by the guiding camera sensor will be small, the fact that the telescope *has* a large aperture means that far fainter stars will be able to be detected. An 8 inch scope will have a gain of ~2 magnitudes over a typical 80 mm guide scope and an 11 inch scope 3 magnitudes. This means that no matter where the guide scope is placed in the allowable field, it is very likely that a guide star can be found and it will rarely be necessary to adjust the guide camera's position.

The fundamental accuracy of a guiding system is ultimately limited by the seeing. Longer wavelengths are less affected by the seeing and this holds true for infrared

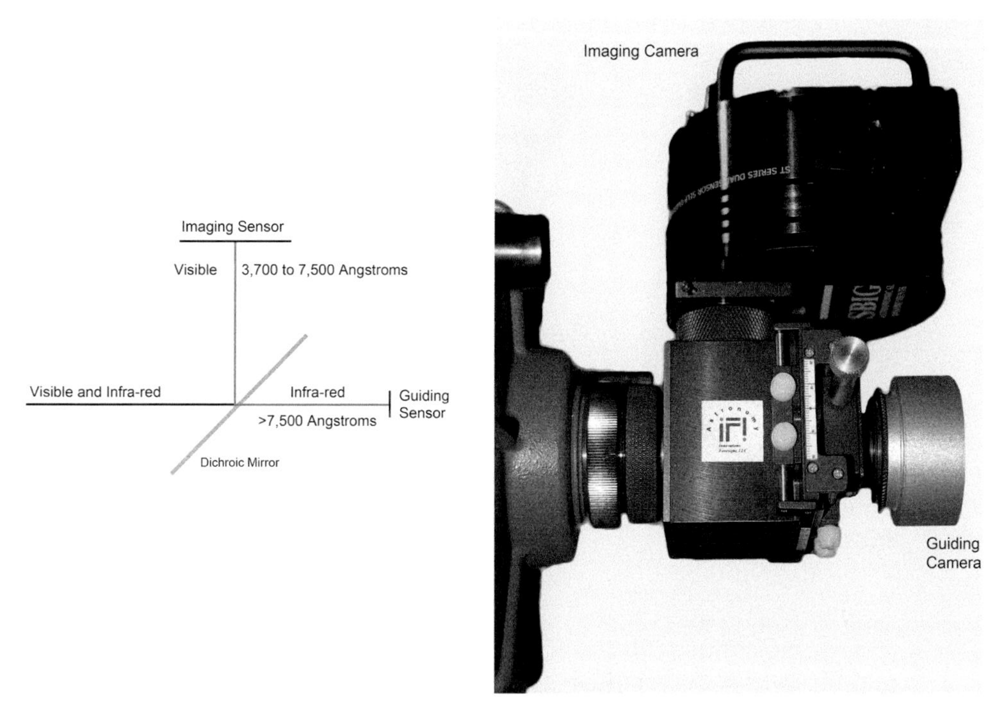

Figure D.4 The Innovations Foresight On-Axis Guider

images, so the fact that this guiding system uses the infrared part of the spectrum for guiding should give a real improvement, with the effect of seeing being reduced by ~23 per cent compared to using visual wavelengths. (This gives one the idea that if an off-axis guide scope system is being used, one might place red or infrared pass filters in front of the guide camera as I have done, as described in Chapter 14, when imaging the Moon.)

Driving the Mount

The computer carrying out the guiding must obviously be able to apply corrections to both the right ascension and declination axes of the telescope. To this end many equatorial mounts are equipped with an autoguiding port. This is a socket for an RJ-12 connector that is usually called an ST-4 port, as it employs the connections originally defined by SBIG for its ST-4 autoguider system and which has now become the de-facto standard. To drive the mount, Dec+, Dec–, RA+ and RA– inputs are grounded as required. Some hardware is required to do this. It may be incorporated within the guiding camera, as discussed below, but otherwise an interface box, such as that supplied by 'Shoestring' in their GP-USB kit, can be used. As its name implies this links to the control computer software using a USB port and provided opto-isolated switches to control the mount to which it is linked by means of a RJ-12 cable.

Figure D.5 Guide camera – Fishcamp Starfish, Starlight Express Lodestar and Orion Starshoot 52064

Some mounts do not have an ST-4 guiding port but, instead, require a serial data control stream from the computer, which can also be used to remotely control the mount from a program such as the *Cartes du Ciel* planetarium program. As virtually no laptop computers are now provided with a serial port, a USB to serial converter may be required.

Guide Cameras

Virtually any small CCD camera could be used, but if one were to buy one specifically what might one look for? It should be monochrome, as a colour version using the same chip will be one third as sensitive. It should have a reasonably large size, so increasing the field of view with a given guide scope and thus providing more candidate guide stars. It should have a high quantum efficiency and be capable of integrating for a few seconds to provide low noise images and thus allow more stars in a given field of view to be used. It should be lightweight to minimise any flex in its mounting to the guide scope and the overall weight of the guide system. Finally, it saves on additional cost and complication if it can include an ST-4 interface port. A nice feature, now common to most guide cameras, is that they are powered through the USB port that is used to control the camera and download its images. This reduces the number of cables that are in use, making for a tidier system.

One excellent guiding camera is the Starlight Express Lodestar, which uses a 6.4 x 4.75 mm Sony IXC429AL CCD chip having 752 × 580, ~8 micron square pixels and an excellent peak quantum efficiency of 65 per cent. It includes an ST-4 compatible interface and is highly regarded. Though not cooled, it can be used to do some useful deep sky imaging as well. It has been the perceived wisdom that CMOS sensors are not as sensitive as CCD chips, but the Micron MT9M001 6.66 x 5.32 mm CMOS chip, which utilises 1280 × 1024, 5.2 micron square pixels and has a peak quantum efficiency of 56 per cent.

It is being used in a number of cameras that are ideally suited for autoguiding purposes. The top flight of these CMOS based cameras, but currently only available for purchase from the USA, is the Fishcamp Starfish camera in cooled or un-cooled versions. This does include an ST-4 interface. At lower cost are the Opticstar AG-130M Coolair and Orion Starshoot 52064. The Opticstar incorporates a fan to help cool the CMOS chip, helping to reduce hot pixels and dark current, while the Orion camera incorporates an ST-4, interface, so an additional GP–USB interface box is not required. Both allow integrations of several seconds. A similar sensor, the MT9M034, is used in the QHY5L-II guide camera, which is also provided with an ST-4 port and, at lower cost than the Lodestar, is now being used in many guide systems.

Most cameras are able to integrate for a few seconds, with a general consensus being that integrations of from two to four seconds are about right for autoguiding: if too short exposures are used, the guiding software might try to 'chase' the seeing that makes a star image move about quickly; if too long, the guiding will not be as accurate. With a few seconds exposure the seeing is averaged out to produce a larger but uniform stellar disc ideal for the guiding software to work with. A somewhat more expensive option is to use one of the entry level cooled CCD cameras, such as the Atik Titan and QHY QHY6, both of which are equipped with an ST-4 guide port. This can give a CCD camera, initially used for imaging, a second lease of life. Cooling may well not be necessary with the short exposures needed for guiding and so there will simply be a single USB2 cable linking the camera to the control computer and the ST-4 guide cable linking it to the ST-4 control port on the mount.

As discussed above, exposure times of a few seconds are best. This implies that the download time of the camera should not be longer than around a second or so, as otherwise this limits the rate at which guide commands can be given to the mount. A guide camera should thus have a USB2 rather than a USB1 serial interface. All modern cameras do use USB2 interfaces which can give download speeds of up to two megapixels per second, thus downloading the ~500 K pixels of a typical guide camera in well under a second. However, older CCD cameras equipped with USB1 or even parallel interfaces will have significantly longer download speeds. This tends to make them less attractive as guide cameras, even though they can often be bought cheaply on the second hand market.

Autoguiding Software

The major CCD imaging packages such as *Maxim DL* include autoguiding facilities, but there are several free stand-alone guiding packages available, such as *Guidedog* and *PHD* (which stands for 'Push Here Dummy'). This latter package is very highly regarded and is very simple to use. One first 'pushes' (i.e. clicks the mouse onto) the camera icon to select the camera type that is being used and which has been connected through one USB port of the control computer. The screen display then shows the imaged star field whilst different integration times are selected. Once having found an integration time such that a suitable star appears and having been prompted to cover the guide scope, four dark frames are taken. (I tend to like the use of a 2 second

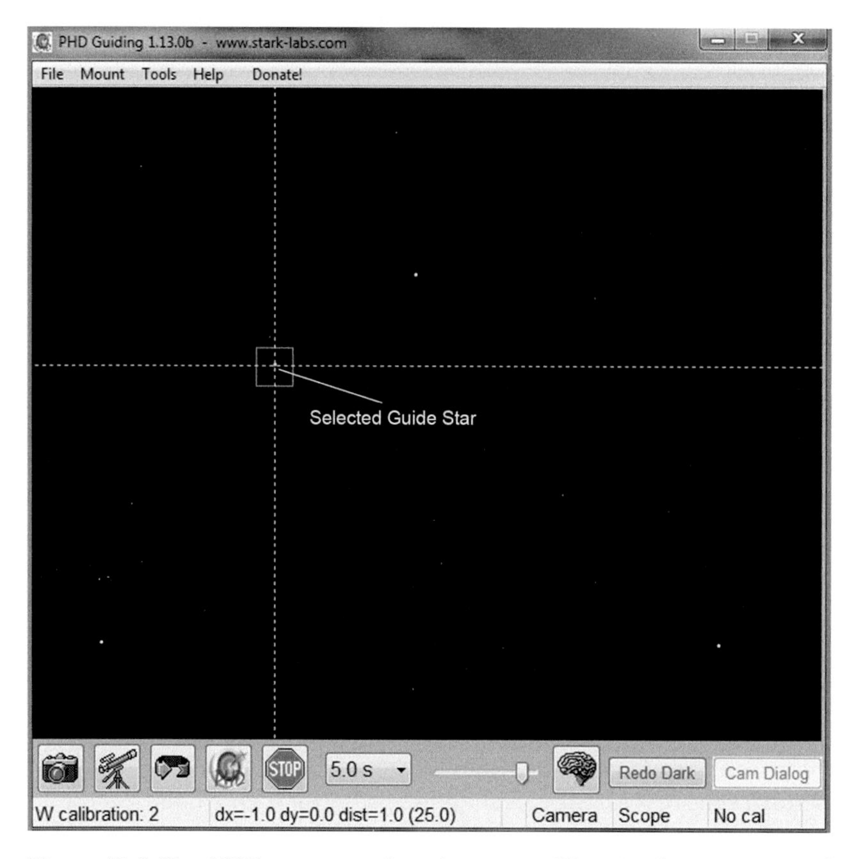

Figure D.6 The *PHD* screen as the telescope calibration has just started

integration time to give 'tighter' control of the mount than with longer exposures and so may have to offset the guide scope somewhat.) These having been taken and averaged, one is prompted to uncover the guide scope. On then 'pushes' a looping icon which initiates the taking and displaying of a continuous sequence of images so that the guide scope can be accurately focused. A suitable star within the field should then be selected. It should not be so bright as to saturate the stellar image but not too faint either otherwise noise can hinder the guiding software. (The program can show a cross section of the star's image so one can tell whether it has saturated the sensor, and if so, reduce the integration period.) When satisfied with the focus and star selection, one finally 'pushes' the *PHD* symbol to initiate the guiding process. It really is that simple! The process begins with an automatic calibration process in which the program sends drive commands the mount and measures in what direction and by how much it moves for a given drive command. When the calibration process, taking perhaps 5 minutes, is complete, imaging can begin. As the way the mount responds to drive commands may vary in different parts of the sky, the calibration procedure should ideally be carried out before each sequence of images is taken.

Sub-pixel Guiding

For use when autoguiding, it is actually better if the stars are not quite in focus, as the guiding software can then carry out what is called 'sub-pixel' guiding. The requirement is to spread a star's light over several adjacent pixels on the camera's sensor. The autoguiding software finds the centroid of the star's image and can then track this to a precision of well within one pixel diameter. As an example, suppose a star's light is spread over a square grid of 4 pixels, with the centre of its image at the precise centre of the grid. Each pixel would then have the same brightness. If the star's image were to move slightly away from this central position towards the right, the two right-hand pixels would receive more light and the two left-hand pixels less. The software, measuring these differences, can then calculate the very slight movement and so apply very precise guide commands to the mount.

Autoguiding Tips

Perhaps surprisingly, it is actually best not to set up the telescope mount perfectly! The ideal situation is when the backlash is removed as far as possible and the guide commands to the mount are always in the same direction. To remove backlash from the right ascension axis, the telescope rather than being perfectly balanced should be made slightly 'east heavy' so that the worm gear is always working on the same side of the worm teeth. This must be done separately for both sides of the meridian flip. The trick to improve declination tracking is to very slightly offset the polar axis from the North Celestial Pole. This causes a slow declination drift in one direction so that all the guide commands will be in the same direction – so again eliminating backlash. The polar alignment error can be quite small so that there will be no problem with field rotation.

Appendix E
Image Calibration

Light, Bias, Dark and Flat Frames

Imaging software refers to four types of 'frame' that relate to a sequence of exposure. The first is a 'light frame', which is a frame taken of the object to be imaged. The second is a 'bias frame', which is a frame that has had zero exposure. It gives a measure of the readout noise or 'bias current' contributed by the chip electronics each time the image data is downloaded into the computer. This may vary slightly with temperature, but at any given temperature will be constant – and hopefully small. The third is a 'dark frame', which is a measure of the total noise due to the bias and dark currents when an exposure is made with the camera shutter closed. It will also expose 'hot pixels' that become more prominent the longer the exposure. If a dark frame has been made with the same exposure time and at the same temperature, this will be subtracted from the light frame images in the first stage of the image processing. In a cooled CCD camera the temperature control electronics will normally stabilise the chip temperature to about half a degree, which means that the dark and bias current noise contribution will remain constant. This means that dark frames (which will be averaged by the imaging software) taken at the start (and perhaps end) of an imaging sequence can be used to correct for all the light images.

As it is obvious that the dark frame will contain the bias current I was, for some time, confused as to why any bias frames need to be taken. It turns out that they are only needed if the light frame exposure time is different from the dark frame exposure – in which case they allow the imaging software to estimate what a dark frame with the same exposure as the light frame would be. This can work as the bias current contribution remains constant and the dark current contribution increases linearly with exposure time. Let us suppose that the light frame exposure was 10 minutes, but there is only a dark frame of exposure 5 minutes along with a bias current frame. (They must all be at the same temperature.) The software will take the 5 minute dark frame and subtract off the bias frame to give only the dark current noise contribution to the dark frame. The noise in this frame is then doubled to give an estimate of the dark current contribution that there would be in a 10 minute dark frame. The noise from the bias frame is then added back in to give an estimate of the total noise in a 10 minute dark frame. The point to make is that if you make dark frames of the same exposure time as your light frames – which is always best – there is no need to take any bias frames.

A number of dark frames, perhaps ~20, will be taken and then averaged in the image processing software. The image acquisition software may well allow a suitable number of dark frames to be taken unattended after the observing session has finished. Assuming that all my exposures are taken at a temperature of –20° Celsius, I can take sequences of dark frames at this temperature and a range of exposures to cover those that I use – typically one to five minutes – at any convenient time.

The fourth type of frame is a 'flat' frame. Particularly if a focal reducer has been used in the imaging chain, the image may suffer from vignetting in the corners of the field. *Adobe Photoshop* has a filter (Filter > Distort > Lens Correction) that allows for vignetting to be corrected in post processing, but a better alternative is to take and average some 'flat' frames before or after the imaging session. A flat frame is simply an image of a uniformly illuminated field. There are several ways of achieving this. One way is to observe a twilight or pre-dawn sky (when the stars are invisible) at an altitude of about 30 degrees towards the south-east or south-west. A second way is to image a matt-white area that is uniformly illuminated. This could be painted on the inside of an observatory dome or a white sheet held in place some distance from the telescope. A third way is to make a light box with a uniformly lit background, while a fourth, similar, approach is to stretch a tight and wrinkle-free sheet or white T-shirt over the telescope aperture and point the telescope at a diffuse light source. The first owner of my 350 mm Maksutov made a telescope aperture cover out of white translucent plastic which is used in just this way. Image processing software will use flat frames to correct for vignetting, but their use will also reduce the effect of dust or hairs on the sensor. Again it is recommended to take quite a number of flat frames, which are averaged by the imaging software. It is important that no part of the image is overexposed. A peak brightness of ~40,000 would work well when taken with a 16-bit camera that has has a sensitivity range of 0 to 65,535.

Appendix F
Practical Aspects of Astroimaging

At Home

Let us assume that the imaging set-up is in the garden and one does not have an observatory. The fact that this is close to the house has two major advantages: firstly, mains power can be used, and secondly, imaging can often be carried out inside in the warm.

Mains power must be used with care and a mains leakage trip placed between the extension lead out to the garden and the mains socket. Mains power could then be used to power all parts of the imaging system, including the mount and dew strips surrounding the telescope objectives or corrector lenses. There is, however, a question regarding providing power to a CCD camera. If the mains power supply is not too well smoothed, it can affect the images, and some have found very low level brightness banding seen in the images which can also be caused by earth loops. Should either of these be the case, then the use of battery power as discussed within the 'Imaging from a Dark-Sky Site' section of this chapter might be a better option.

If the garden, as in the case of my own, cannot be easily accessed from the road, then it is possible to leave the mount permanently in position – suitably protected from the elements of course. This has the great advantage that, once aligned on the North Celestial Pole, it should not be necessary to realign before each observing session. The control unit of the Astro-Physics mount that I now use can be easily removed from the mount so that there is no possibility that it could be damaged by condensation when not in use. As a next step, a pier can be placed in the garden so that only the equatorial head need be mounted upon it at the start of each observing session. This will normally keep a reasonable alignment if the connecting bolts are not too loose. In order to observe in the west, I have to use a Losmandy GM8 which stands on concrete paving in front of my house and have set down some markings so that I can precisely locate the three tripod legs, which are kept to the right length to ensure that the tripod platform is horizontal.

Figure F.1 Focusing an imaging camera using a Bahtinov mask

Observing from Inside the House

This has the obvious advantage that, once having set up the mount, telescopes and cameras, it enables one to keep warm during the imaging session. But there are several considerations to take care of. The majority of cameras, both for imaging and guiding, are controlled using USB connections that have relatively short leads. It is possible to buy powered USB extension cables to extend these into the house, or even better, to have a USB hub adjacent to the telescope so that only one powered USB extension cable is required.

Assuming that the control laptop or computer will be used inside, there can be a problem when focusing the imaging and guide camera, unless one has remotely controlled focusers. Initially I located the laptop adjacent to the telescopes so that I could observe the images produced by the imaging and guide programs and so be able to focus them. Without disconnecting the cables I then had to carry the laptop inside. This was not always successful. To avoid this problem, I simply bought a 30-ft long VGA extension cable and a very cheap monitor to set beside the telescopes so that, linked to the external VGA connector on the laptop, I could focus the cameras while keeping the laptop inside.

Remotely Controlling the Mount

When initiating an imaging session I will align and then synchronise the mount on a bright star which would be used when focusing both the imaging and guide cameras using a Bahtinov mask as seen in Figure F.1. Then one can then slew to the object to be imaged and centre it in the field of view of the imaging camera. If guiding is to be used, one would then select a suitable guide star, calibrate the guiding commands, as described in Appendix D and initiate autoguiding. (It is best to calibrate the guide commands where the mount is to be tracking as these might need to change in different regions of the sky.)

If one is only going to image one object during an observing session, then the mount need not be remotely controlled, but if not, it is nice to be able to command the mount from inside. Some mounts will have a RS232 serial port to accept remote commands. This is a problem, as laptops do not now have serial ports, but USB to serial converters can be purchased to overcome this problem. Many imaging programs allow for telescope control as do planetarium programs such as *Cartes du Ciel*, *Stellarium* and *Sky Safari Plus*, which I use to control my Astro-Physics mount from an Apple Mac computer. To avoid a cable connection, I have a Southern Stars SkyFi unit. This provides an RS232 (a USB port is also included) connection to the mount and sets up a local WiFi hub. This WiFi channel can be selected by the *Sky Safari* program and then one can drive the mount to any visible celestial object by simply clicking on it and then making slow speed pointing corrections as required.

Imaging from a Dark-Sky Site

Mounts

The mount that is used at one's home location may well be portable so that it can be taken to a dark-sky location, but if not, it can be worth considering an alt-az mount as these are so much easier to set-up. The iOptron MiniTower can easily support telescopes up to 102 mm refractors or 150 mm Newtonians, Maksutovs or Schmidt–Cassegrains. Once having set the tripod platform horizontal using a spirit level, the alt-az head can be very accurately levelled with three further fine adjusters at its base. A GPS unit downloads the precise location and time. The head needs to aligned due south with the telescope vertical but this does not need to be too accurate. The mount is then commanded to slew to a bright star or planet. If this is not in the centre of the field of view, then it is easy to see in which direction to drive the mount in azimuth to bring the object into the field. If the head is accurately horizontal, only very slight altitude corrections will be required. If the object that one wishes to image, say Jupiter, was chosen to slew to, then setting up the mount takes essentially no time apart from setting the tripod platform and the head horizontal. If using an equatorial mount, it might be well worth purchasing a QHY PoleMaster, described in Appendix B, which can be used to accurately align it in just a few minutes.

Battery Power

The majority of mounts, cooling systems for cameras, dew tapes and other ancillary components run on 12 volts DC. There are quite a number of options. SkyWatcher and Orion provide 'Power Tanks', having 7 or 17 ampere-hour capacities and which also include lamps. Two 12 volt 'car-type' sockets are provided. One of these power tanks is certainly useful, but the cost per ampere-hour is quite high. Many companies now provide 'jump-start' batteries that usually have one 12 volt socket, which can work out significantly cheaper. Probably the lowest cost per ampere-hour can be obtained with the gel-cell batteries used for golf buggies. These have a capacity of 22 ampere-hours but will need to be provided with a 12 volt socket or a direct connection to a suitable plug for, as an example, to supply power to a mount. These gel-cell batteries have to be carefully charged with a suitable charger. One of the best is the NOCO G1100 'Genius wicked smart charger', which can charge both 6 and 12 volt gel-cell batteries but can also charge 12 volt lead-acid car batteries providing a maximum current of 1.1 amps. It is interesting that iOptron sell a 'PowerWeight' counterweight battery pack that provides an 8 ampere-hour battery that also acts as a 3.2 kilogram counterweight.

The latest batteries are based on Lithium Polymer technology and manufactured by Tracer High Performance Batteries. They weigh just a quarter of equivalent capacity lead-acid base batteries and have a very flat discharge characteristic. Their capacities range from 4 to 22 ampere-hours. An LED display shows how much charge is remaining and a charger is supplied with each battery. Their only disadvantage is their rather high cost per ampere-hour.

It is likely that a laptop will be used to acquire the images and perhaps control the autoguiding system. Their batteries may well not last for a full night's imaging session. A solution is to purchase additional battery packs, but the changeover from one to another may not be convenient. An alternative solution is to use a high capacity 12 volt battery and 1000 or 1500 watt DC to AC inverter. If there are other observers and imagers nearby, such as at a star party, the light from a laptop screen can be a real problem. Some programs, such as *Astro Photography Tool* purposely use a red screen for their display, but it is a good idea to cover the screen of the laptop with a red gel. Some imagers will use an open sided box in which to locate the laptop to limit any light leakage. Red torches should of course be used and it is usually forbidden to use a green laser pointer.

Appendix G
A Survey of Astronomical Websites and Instructional DVDs

To Aid Planning

Astronomy Tools Field of View calculator: https://astronomy.tools/calculators/field_of_view/

Virtual Moon Atlas: https://sourceforge.net/projects/virtualmoon/

Stellarium Planetarium program: www.stellarium.org/en_GB/

Sky Safari Planetarium programs: http://skysafariastronomy.com/products/skysafari/

Cartes du Ciel: www.ap-i.net/skychart/en/start

SkyX Pro: www.bisque.com/sc/shops/store/TheSkyX+Editions/default.aspx?PageIndex=0

SkyMap Pro 11: www.skymap.com/smpro_main.htm

Image Capture

BackyardEOS for Canon DSLRs: https://www.otelescope.com/index.php?/home/&sku=OTL-BYE-T

Astro Photography Tool for Canon DSLRs: www.astropix.com/wp/2015/12/11/apt-astro-photography-tool-v3-0-is-here/

BackyardNikon for Nikon DSLRs: www.otelescope.com/index.php?/home/&page=byn

Digicamcontrol for both Canon and Nikon DSLRs: www.digicamcontrol.com

EZcap for QHY CCD cameras: www.qhyccd.com/Download.html

CCDops for SBIG cameras: www.sbig.com/support/software/

FireCapture for many planetary webcams: www.firecapture.de/

IC Capture for Imaging Source webcams: https://www.theimagingsource.com/products/software/

FlyCapture for Point Grey webcams: https://www.ptgrey.com/support/downloads

Maxim DL for many CCD cameras: www.cyanogen.com/maxim_main.php

Nebulosity for many CCD cameras: www.stark-labs.com/nebulosity.html

ImagesPlus for many CCD cameras: www.mlunsold.com/ILCameraControl.html

Autoguiding Software

PHD Guiding: www.stark-labs.com/phdguiding.html
GuideDog: http://barkosoftware.com/GuideDog/index.html
MetaGuide: www.astrogeeks.com/Bliss/MetaGuide/index.html

Planetary Video Processing

Registax: www.astronomie.be/registax/
AutoStakkert! 2: www.autostakkert.com/wp/

Combining Sub-Exposures

Deep Sky Stacker: http://deepskystacker.free.fr/english/index.html
StarStax: www.markus-enzweiler.de/StarStaX/StarStaX.html

Star Trails Stacking

StarStax: www.markus-enzweiler.de/StarStaX/StarStaX.html
Startrails: www.startrails.de/html/software.html
Starmax: http://ggrillot.free.fr/astro/starmaxEng.html

General Image Processing

Adobe Photoshop: www.adobe.com/uk/creativecloud/catalog/desktop.html
GIMP: www.gimp.org/
[Note: Version 2.10, under final development at the time of writing, will support 16-bit processing and will then become more suitable for astronomical image processing.]
Astra Image deconvolution sharpening tools: www.phasespace.com.au/

Photoshop Actions

Noel Carboni's Astronomy Tools: www.prodigitalsoftware.com/Astronomy_Tools_For_Full_Version.html
Annie's Astro Actions: www.eprisephoto.com/astro-actions
GradientXterminator: www.rc-astro.com/resources/GradientXTerminator/

Astronomical Imaging Packages

Maxim DL: www.cyanogen.com/maxim_main.php
PixInsight: https://pixinsight.com/

Astroart 6.0: www.msb-astroart.com/
StarTools: www.startools.org/
IRIS: www.astrosurf.com/buil/us/iris/iris.htm
FITS Liberator: www.spacetelescope.org/projects/fits_liberator/

DVD Astronomy Tutorials

By Jerry Lodriguss: www.astropix.com/agda/agda.html
A Beginner's Guide to DSLR Astrophotography
A Guide to Astrophotography
A guide to DSLR Planetary Imaging
Astrophotographer's Guide to the Deep Sky
Photoshop for Astrophotographers

By Damian Peach:
High Resolution Astrophotography Part I
High Resolution Astrophotography Part II
UK: www.widescreen-centre.co.uk/astronomy-books-dvds-and-software/astrophotography/high-resolution-astrophotography-dvd-part-i
USA: http://starizona.com/acb/High-Resolution-Astrophotography-by-Damian-Peach-P3708C0.aspx
Europe: www.astromarket.org/accessoires/star-chart---dvd/d-peach-hr-astrophotography-dvd-english
Canada: www.khanscope.com
Australia: www.myastroshop.com.au/

Index

Printed in the United States
By Bookmasters